A Common Language for Electrical Engineering

Lone Pine Writings

A COMMON LANGUAGE FOR ELECTRICAL ENGINEERING

A Common Language for Electrical Engineering

Lone Pine Writings

Eric P. Dollard

Published by
A&P Electronic Media
Spokane, Washington

A COMMON LANGUAGE FOR ELECTRICAL ENGINEERING

Cover Design: Aaron Murakami
Editors: Peter Lindemann, Jeff Moe & Aaron Murakami

Cover Image: shutterstock.com #108304397

Author Copyright © 2011–2015 Eric P. Dollard
Publisher Copyright © 2013–2015 A&P Electronic Media

All Rights Reserved, Worldwide. No part of this publication may be reproduced, stored in an electronic retrieval system, or transmitted in any form, or by any means, without the prior, written permission of the copyright holders or the publisher. Unauthorized copying of this digital file is a criminal offense and prohibited by International Law.

Digital and Print Edition Published by:

A&P Electronic Media
PO Box 10029
Spokane, WA 99209
http://emediapress.com

Digital and Print Version 1.10 – Release Date: December, 2015
Digital Version 1.00 - Release Date: October, 2013

Digital version available at http://lonepinewritings.com

ISBN-13: 978-1518815935
ISBN-10: 1518815936

Table of Contents

Foreword — 1

Transmission 1 — 3
Energy Defined in Terms of Electrical Engineering

Transmission 2 — 6
Rates of Change

Transmission 3 — 15
The Planck Revisited

Transmission 4 — 21
The Time Tunnel

Transmission 5 — 28
Space, the Final Frontier

Transmission 6 — 32
Dimensional Meanings

Transmission 7 — 62
In Space, Heaviside

Transmission 8 — 79
Experimental Cubic Volumes

Transmission 9 — 87
Inductance and Capacitance

Transmission 10   112
The "Telegraph Equation"

Conclusion   131

List of Symbols   133

Recommended Reading   134

Glossary   135

Foreword

This book, and its content, developed out of the general frustration of the author when trying to teach others about his work in electrical engineering.

This collection of essays started appearing in discussion threads on the Energetic Forum around 2011. At the time, the author, Eric Dollard was living in his car near Lone Pine, California. Originally, Eric wrote the material out on paper and sent it to a colleague who transcribed the material and posted it in the forums under the pseudonym "T-REX". As of the time of this publication, all of these posts are still available on the internet.

Each letter was called a "transmission" and contained information on specific electrical engineering terms and how they are to be used. The original format of the material is retained in this edition of the book.

The phenomena we call "electricity" is a dynamic, but artificial presentation of the Natural World, and because of this, its behavior follows specific rules. Understanding these specific behaviors is the key to engineering this phenomena, but developing a common language with which to describe these behaviors is the key to teaching others these engineering skills.

The purpose of this book is to provide clarity for the electrical engineering community regarding the use of common terms for electrical units. The last attempt to standardize this language was made by Oliver Heaviside over 100 years ago, and his effort was met by censure from the Royal Society of London. It is hoped that the release of this book will be met with a more enlightened response.

Peter A. Lindemann, D.Sc.
Editor

Transmission 1

Energy Defined in Terms of Electrical Engineering

Now, why are we all gathered here? The focus is on "Energy", but what is energy, and why is it so important to everyone anyway? It seems somewhat obsessive. The definition of it is the ability to do work, but vernacular has broadened this, so now energy can mean almost anything. This must stop.

Energy, in its most arch-typical form, is embodied in the phenomenon of Electricity, but what is Electricity? Now our wheels even more stuck in the mud! But we have important clues, namely that of polarity, not plus or minus so much but more like male or female. This thought follows from Goethe to Tesla and Steinmetz. Thus Electricity, in order to manifest, a UNION must develop. This is the union of the "male", or projective, and "multiplied by" the "female", or receptive. Hereby, the male is the dielectric field in counterspace (of per centimeters), and the magnetic field or female in space (of centimeters squared). Space in cm squared is what you pay for in "real estate", counterspace in per cm is the space between the lines on a ruler, or between molecules in a crystal.

For the Electricity extant between a pair of wires in your lamp cord, the closer the wires, the more capacitance, and thus the more Dielectricity. Conversely, for the same cord, the farther apart the wires, the more inductance and thus the more Magnetism. Therefore it is seen that the smaller the space (the more counterspace) the more Dielectricity that can be stored, and conversely the larger the space between the wires (the more real estate) the more Magnetism that can be stored. Very simple, do not let your mind make it any more complicated than that!

Now let us reach out for a few quantitative relations: The product (line, cross, or dot-unrestricted) of the total amount of Dielectricity multiplied by the total amount of Magnetism (when both are in union) gives the total quantity of Electricity. We will call this quantity of Electricity the letter "Q" and name this "The Planck" after Max Planck. For the Einstein dimensions of the Planck are Energy-Time, but let us not think backwards-ass.

Saying this in engineer's lingo, the quantity of Electricity, Q, is given as Watt-Seconds-Seconds, or Watt Seconds squared. Now, in one foot of lamp cord, bounded between the wires, I have, say, one million Plancks of electric induction. The frequency is 60 cycles (377 radians) per second. Thusly the quantity of Plancks, Q, is being produced or consumed at a time rate of 377 radians per second, or in other words, Plancks per second. Hence the time rate of variation of the quantity of electric induction hereby gives; Watt seconds squared per second, or dividing out, gives Watt seconds. But Watt seconds is the dimensions of energy. Well golly-gee Mr. Wizard, we have defined energy! And hereby power is defined as the time rate of the production or consumption of the electric induction, Q, divided by time, which gives Watts.

It is that simple. So push the "Erase Button" on your head for two notions: Energy is the product of mass times the velocity of light squared, erased? Next, Electricity is the flow of electrons in the wire, erased? Good!

Transmission 2

Rates of Change

In the previous transmission it was shown that the electric induction, bound between the wires of a lamp cord, was the union of two distinct fields of induction, the dielectric in counterspace, and the magnetic in space. These fields consist of discrete lines of force. Thus these lines exist as individual units or quanta of inductive force. Both fields exert mechanical force upon the bounding system of so-called "conductors". These mechanical forces, those of the dielectric, and those of the magnetic, exert actions so as to increase their coefficients of induction, that is the dielectric "capacitance", and the magnetic "inductance" are increased.

Hereby, the dielectric field draws the conductors nearer to each other, increasing the counterspace. Conversely the magnetic field pushes the conductors away from each other, increasing the space. Hereby we may say that the dielectric field is contractive, and the magnetic field is expansive. Hence the resulting electric field of the union produces a resultant force upon the bounding conductors. This resultant force thus may be expansive, null, or contractive, depending upon the relative densities of the dielectric and the magnetic force fields respectively.

So now our previous discourse upon these matters brings important questions to mind that heretofore remained unanswered: First, how big is a Planck? In other words, how many Plancks per one second (unit time) equal one Watt-second (unit energy)?

Second, what ratio of dielectric field density to magnetic field density results in the contractive force just balanced against the expansive force, thereby canceling any mechanical forces upon the bounding conductors? Who can solve these important questions?

The following references may help answer these questions:

1. Electro-magnetic Theory Vol. 1 by Oliver Heaviside.
2. Impulses, Waves and Discharges by Charles P. Steinmetz.
3. Electricity and Matter by J.J. Thomson.
4. Recent Researches Into Electricity by J.J. Thomson
5. Discharges In Windings by E.P. Dollard
6. Occult Ether Physics by William Lyne

Rates of Change

So far I think the concept of space and counterspace in its basic form is established. Counterspace, as that space between the lines on a ruler is an apt descriptive analog. A ruler divided in millimeters has less counterspace than a ruler divided in nanometers. This is obvious. (Think in terms of capillary action) Also, the Planck is our undivided fundamental quantity of electricity and all else will be a development of the Planck.

What needs to be focused upon at this point in writing is the concept of VARIATION WITH RESPECT TO TIME, that is the dimension of per second. This also is known as the "Time Derivative", delta X over delta T, in the Newton-Leibniz infinitesimal calculus. Let us say the rate of change with respect to time.

Everyone's intimate mechanical relationship with their automotive apparatus render it useful tool for analogy. Various phenomenon make their appearance, somewhat ghostlike, during the process of variation with respect to time.

Let us take the dimensional relation of velocity, that is, the ratio of the dimension of length (space) to the dimension of time. This is the velocity V, let us say specifically miles per hour. Now we know that if the dimension of mass, m, is moving at a given velocity, *V*, that is, the "weight" of your body in the auto, in CONSTANT motion at the speed limit let's say, no forces or perceptible sensation is imparted to your body. In other words you feel sitting in the car seat the same as you feel in the seat in front of the television set (so the auto is like a rolling TV). So long as the velocity remains unchanged nothing is experienced or felt. Now a deer jumps in front of your auto, you slam on the brakes and miss it, now you are moving at 1/10th of the speed limit. During the interval in time in which the velocity varied with respect to time, that is miles per hour per second, or the ratio of the velocity *V* to the time interval *t*, your physical body experienced a powerful force pushing you forward. The more quickly the auto changed speed the more this force impacted your body. So we can say that this force *F* is given as the ratio of velocity to time, for any unit mass of your body. That is, the force *F* equals your body's mass times the ratio of the velocity to the time interval of the velocity's variation with respect to time.

So back to the Planck, that quantity of electrical induction, *Q*. So long as there is no variation of the electrification, that is so long as it is static, no other phenomenon manifests. Just as with the mass of your body and the velocity of its motion in space, so long as there's no variation no sensation of force is experienced by your body. Completing the analogy the concept of energy is then entirely analogous to that force you felt when you hit the brakes. Hence they are both phantom like derivatives of things that you otherwise can perceive as real.

So now we are getting a further "feel" that what we call energy is not really that primary phenomenon that the Einstein would like us to believe, but in reality is only a secondary derivative of some more concrete phenomenon, or ghost associated with something physically tangible or real. Now the idea that energy can be "created" or made to go away can now be brought to mind.

Continuing with Variation of a Dimension with Respect to Time

Continuing with the concept of the variation of a quantity (or dimension) with respect to time (another dimension). We may say then we are talking about a RATIO of a physical dimension to a metrical dimension. Previously given, the ratio of a physical dimension, the Planck Q to a metrical dimension the time t gives then the dimensional relation of energy W. Then from the Newton-Leibniz concept we say delta Q over delta t equals W, that is, the first order time derivative of electrification Q equals the energy W. Now Einstein says the inverse, and that is, the time integral of W, the energy, over time interval t' to t" is the electrification Q. This is to say Q is the PRODUCT of the energy W and the time interval T. W times T equals Q. This is backward-ass, thereby occluding the interrelationships of these three distinct relationships.

Further, hit your erase button on the gibberish of 1, 2, or 3 dimensional space, there is only ONE DIMENSION OF SPACE – SPACE! Coordinates are NOT dimensions. Example, the volume of a cylinder can be expressed in TWO terms, height and circumference. So where is the third "dimension", erased?

Continuing then it has been given that the total electrification Q is the union, or product, of the total dielectric induction Psi, and the total magnetic induction Phi. In other words, the dimensional relationship Q, the total electrification, is the product of the dimension of total dielectric induction Psi, and the dimension of total magnetic induction Phi. Hence we have FOUR primary dimensions in electrical engineering.

These are:

1) Time
2) Space
3) Dielectricity
4) Magnetism

Every other relation, quantity, or expression, Volt, Amp, Ohm, etc. is derived from these FOUR dimensions. Time and Space are the metrical dimensions; Dielectricity and Magnetism are the physical dimensions. It is that basic!

Time Variance and Its Products

Continuing on the concept change with respect to time, the total electrification Q in Plancks is a resultant of the union, or PRODUCT, of a pair of inductions, the total dielectric induction, Psi and the total magnetic induction, Phi.

Now let us deal with these two inductions individually. Variation of the total dielectric induction Psi with respect to time t, that is, the RATIO of the dimension of dielectric induction to the dimension of time, dielectric induction over time, or the time rate with which the dielectric induction is produced or consumed, the DISPLACEMENT CURRENT in amperes I. Let us call this Maxwell's Law. PSI over t equals current I. This makes sense since the charge in a battery is given as Ampere-Hours, that is I times t equals Psi.

Hence we have arrived at a new dimensional relationship, the "current" in Amperes.

Analogously, we have the variation of the total magnetic induction Phi with respect to time t, that is, the RATIO of the dimension of magnetic induction to the dimension of time, magnetic induction over time, or the time rate with which the magnetism is produced or consumed, is the electro-motive force in Volts. Phi over t equals the voltage E. We know this as the "Faraday law". Hence we have arrived at a new complimentary dimensional relationship, the E.M.F. in Volts.

Taking this one step further, consider the ratio of the variation of the magnetic induction Phi with respect to time t, E, to the variation of the dielectric induction Psi with respect to time t, I. Since the dimension of time appears on both top & bottom of the ratio this dimension cancels leaving simply the ratio of Phi to Psi. Hereby the RATIO of the total magnetic induction Phi to the total dielectric induction Psi gives the dimensional relation of impedance, Z, in Ohms. This can be arrived at somewhat differently. Since the variation of magnetism with respect to time is the electro-motive force, E, and since the variation of dielectricity with respect to time is the displacement current, I, then the RATIO of the electro-motive force, E, to the displacement current, I, give the impedance, Z, in Ohms. That is, the ratio of E to I is Z in Ohms.

Inversely, from the standpoint of dielectricity rather than magnetism the RATIO of the total dielectric induction, Psi, to the total magnetic induction, Phi, that is the ratio of dielectricity to magnetism is the "Admittance", Y, in Siemens. Hence the ratio of the displacement current I to the electro-motive force E gives the Admittance, Y, in Siemens.

Hereby we have arrived at four new distinctive dimensional relationships:

1. Displacement current in amperes, **I**
2. Electro-motive force in volts, **E**
3. The impedance in ohms, **Z**
4. The admittance in Siemens, **Y**

Transmission 3

The Planck Revisited

Continuing with the conceptualization of RATES OF CHANGE with respect to time and the interactions that arise.

In summary it has been given that variation of a quantity of dielectric induction, Psi, with respect to time is the DISPLACEMENT CURRENT, in amperes, I. We will call this Maxwell's Law of Dielectric Induction after its discoverer James Maxwell.

Also given is that the variation of a quantity of magnetic induction, Phi, with respect to time is the ELECTRO-MOTIVE FORCE in volts, E. We will call this Faraday's Law of Electro-Magnetic Induction, after its discoverer Michael Faraday.

It should be noted that displacement currents flow thru the insulation (dielectric). It is NOT the familiar conduction current of the electronic ideologies.

Likewise the electro-motive force is a result of the conductor (metallic). It is NOT the electro-static potential of the dielectric field. A conjugate relation exists here, the "insulator" and the "conductor". So now we have TWO distinct "volts" and TWO distinct "amps", hence E and I are seen to have dual definitions. The ampere, *I*, may be a displacement current, or may be a conduction current. Likewise the volts, *E*, may be an electro-motive force, or it may be an electro-static potential. These distinctions are important and misunderstanding rests here.

The Maxwell-Thomson concept of electric induction, and the aether which engenders this induction, considers the dielectric lines of force, and the magnetic lines of force, as CONCRETE PHYSICAL REALITIES. (Read Electricity and Matter by J.J. Thomson, and also read Theory of Light and Color by Babbitt, the Un-sterilized version).

These lines can be considered "tubes of force" a hydro-dynamical vortex tube of sorts. Here we find the "hydro-dynamical model of the aether" as given by James Clerk Maxwell. Understanding of this sort has been buried by the relativists and quantum car mechanics. From the initial concept of Faraday, thru the theoretical reasoning of Maxwell, into the experimentalist like Crookes and J.J. Thomson, it gave an ENGINEERABLE CONCEPT of the primordial aether. Finally Nikola Tesla, Oliver Heaviside, and Charles Steinmetz turned this into today's electrical technology. The roots of Edison sprang to life.

So what may aether be? Consider what are called the "states of matter".

1. SOLID
2. LIQUID
3. GAS
4. PLASMA
5. AETHER

Hence, the five distinct states of matter.

Electricity is embodied in the aetheric state of matter, or "proto-matter". Electricity is aether in a state of dynamic polarization; magnetism is aether in motion, dielectricity is aether under stress or strain. The motions and strains of the aether give rise to electrification. Phi times Psi gives Q.

In defining the hydro-dynamical tubes of force as concrete realities, a distinct phenomenon taking place with the aether, the constitution of the Planck sticks its snout out of the sand. The tubes of force are discrete, fiber-like, quanta as some would say. Experiments by J.J. Thomson indicate this. Lines of force are a quantum phenomenon, distinct concrete entities.

Further, we have the idea of "Planck's Constant", any variation in the total density of electric induction Q, in Planck's, cannot vary continuously but must exhibit its variation in discontinuous, or discrete steps. Hence a distinct quanta Q. We may infer that the union, or CROSS PRODUCT, of a single tube of DIELECTRIC induction, with a single tube of MAGNETIC induction, gives birth to a single unit of ELECTRIFICATION, Q. This idea embodies the concept of the photon, a QUANTUM UNIT of electromagnetic induction. Also consider the J.J. Thomson concept of the "electron" (his own discovery). Thomson considered the electron the terminal end of one unit line of dielectric induction. One tube, one electron. So then, how big is a unit Planck, the quantum unit of electric induction, Q?

Let us summarize the knowledge we have gained from what has been given to this point.

The basic engineering dimensional relationships are hereby,

$$\Psi\Phi = Q$$

The undivided quantity of the total electrification, **"Planck"**,

$$\frac{Q}{t} = W$$

The time rate of the production or consumption of this electrification, **"Joule"** (energy),

PSI, Ψ

The total dielectric induction, or the ratio of the total electric induction Q, to the total magnetic induction which is embodied in this electric induction.

This is the "**Coulomb**" (charge),

PHI, Φ

The total magnetic induction, or the ratio of the total electric induction, Q, to the total dielectric induction, Psi, which is embodied within this electric induction.

This is the "**Weber**" (induction),

$$\frac{\Phi}{t} = E$$

The electromotive force which results from the production or consumption of the total magnetic induction Phi.

The unit is the "**Volt**",

$$\frac{\Psi}{t} = I$$

The displacement current which results from the production or consumption of total dielectric induction Psi. The unit is the "**Ampere**".

Group one consists of: Q, Plancks, Psi, Ψ, Coulombs, and Phi, Φ, Webers. Group two consists of: W, Joules, E, Volts, and I, Amperes.

Group one represent PRIMARY quantities, whereas group two represent REACTIONS by the primary quantities to their variation in quantity with respect to Time.

Transmission 4

The Time Tunnel

Oliver Heaviside once stated that "The law of continuity of energy" is maintained when the energy existent at one time disappears but reappears at another time. In H.G. Wells' the "Time Machine" the professor argues with the doctor saying that his time device (which just vanished) is still at the same spot in the room, but the doctor can't see it because it is there, yes, but in another time. It moved through the dimension of time.

The motion of electricity in time has been the primary topic so far.

Given are the three basic relations,

Plancks per second gives Joules

$$\frac{Q}{t} = W \qquad \text{Joules}$$

Coulombs per second gives Amperes

$$\frac{\Psi}{t} = I \qquad \text{Amperes}$$

Webers per second gives Volts

$$\frac{\Phi}{t} = E \qquad \text{Volts}$$

Joules, Amperes or Volts, are SECONDARY reactions in response to variation of our known PRIMARY dimensions, the total electrification in Plancks, the total dielectrification in Coulombs, the total magnetization in Webers.

In terms established by Oliver Heaviside, the Volts of E.M.F. are a MAGNETIC REACTANCE, and the Amperes of displacement are a DIELECTRIC SUSCEPTANCE. The degree to which the reactance and the susceptance manifest is proportional to the time rate of variation, that is, per second.

Here the dimension of time is not seconds, it is PER SECONDS. Somewhat like counterspace, but this is NOT COUNTER TIME. One instance of a per-time arrangement is cycles per second. This is known as the FREQUENCY, F, in cycles per second. Here is a dimensional relation of per second, frequency, F. This frequency F represents only a rotational (alternating) cycle, and thereby is only a partial frequency. Also existent is a cycle of geometric progression or regression. This "frequency" is given in decibels per second. Hence our general frequency is given as decibel-cycle per second, or for Newton-Leibniz methodology, it is Neper-Radian per second, v.

Hereby,

1. $v\,Q$ equals W, Joule or Planck per second.
2. v PSI equals I, Ampere, or Coulomb per second.
3. v PHI equals E, Volt, or Weber per second.

The dimensional factor, v, in per second we will call the Heaviside "Time Operator" This time operator describes the variation with respect to time as a "**versor operator**".

E and I are NOT necessarily time coincident, but one may lag or lead the other. Cause and effect become separated by what is known as HYSTERESIS. It can be said that E and I exist in different "time frames". This subject rapidly accelerates into a Bach type reality and is much too complex for now.

Ratios and Products Continued

Up to this point has covered the concepts of dimensional relationships known as ratios, dimension "one" per dimension "two". If dimension two is that of time, the ratio becomes a time rate or time derivative, from the Newton-Leibniz concept. This ratio is known as a first order time derivative, or differential equation. The dimension of time is in PER SECOND and this may be called a FREQUENCY, v, in NEPER-RADIANS per SECOND. But let us not plunge these depths quite yet.

So what about products, the union of dimensional relation "one" by the dimensional relation "two", the product of one and two? Given thus far is the product of the magnetic induction, Phi, and of the dielectric induction, Psi, giving forth the total electric induction, Q. The product of magnetism united with dielectricity gives rise to the total electrification of the aether. Psi times Phi.

But consider the union of the law of magnetic and dielectric induction. Faradays' times Maxwell's. Thus the union, or product, of the electro-motive force, E, in volts with the displacement current, I, in amperes. E times I. Here, specifically, is the product of the dimensional relation WEBERS per SECOND, and the dimensional relation COULOMBS per SECOND.

The resultant relationship is hereby,

WEBER-COULOMB per SECOND SQUARED

$$\frac{\Phi}{t} \cdot \frac{\Psi}{t} = \frac{\Phi\Psi}{t^2}$$

But it has been given that,

WEBER-COULOMB equals PLANCK

$$\Phi\Psi = Q \qquad \text{Planck}$$

Thus the dimensional resultant of the union of the pair of dimensional laws is PLANCKS per SECOND SQUARED. We will call this the electrical ACTIVITY, also known as the electrical power, P.

Hence the dimensional relation,

PLANCKS PER SECOND SQUARED equals WATTS

$$\frac{Q}{t^2} = P \qquad \text{Watts}$$

E times I equals P, Volts times amperes equals watts. However it has been given that the energy, W, in Joules is dimensionally the time rate of the total electric induction, Q in Plancks.

That is,

JOULES PER SECOND equals WATTS

$$\frac{W}{t} = P \qquad \text{Watts}$$

That is, the electrical activity in Watts represents the variation of the total energy of the electric field, this energy itself resulting from the variation of the total electric field of induction.

This is not unlike the situation in the automobile. No forces appear with the first order time derivative of miles per hour, but manifest in direct proportion to the second order time derivative.

That is,

MILES per HOUR per SECOND
or
MILES per HOUR-SECOND

$$\frac{\text{mile}}{\text{hour} \cdot \text{second}} = \frac{l}{t^2}$$

Hence, there is a distinct similarity between the dimensional relation for mechanical reactive force and the dimensional relation for electric activity.

That is,

PLANCK per SECOND-SECOND

$$\frac{Q}{t^2} = P \qquad\qquad \text{Watts}$$

The Watt of electrical power, P.

Transmission 5

Space, the Final Frontier

As a dimension, space is distinct from the dimension of time, and is devoid of any physical dimension. It is hereby eternal, and empty. Like time, space is a metrical dimension, it exists to quantify. Bounded space can define a volume, area, distance, span, or density.

It is customary to consider space boundaries as a CUBIC, or third degree set of coordinates. The three coordinates are length, width, and height, taken from a corner of the cube. Think of a sugar cube, the sugar is the space and the corners define the boundaries. These three coordinates, length, width, and height are WRONGLY known as the three dimensions of space. This is a major mind virus and is hard to erase.

There is only one dimension of space, SPACE, a metrical dimension. Any number of coordinates in any number of geometries can serve to define the boundaries of said space. The use of the cubic three is habitual.

The dimension of space is considered to exist in degrees, or powers of a unit space dimension, here centimeters, l (lowercase L). So we can say cubic centimeters, or square centimeters, etc.

A COMMON LANGUAGE FOR ELECTRICAL ENGINEERING

Hereby, on a cm basis,

$$l^1 \quad \text{distance}$$

$$l^2 \quad \text{area}$$

$$l^3 \quad \text{volume}$$

and

$$l^{-1} \quad \text{span}$$

$$l^{-2} \quad \text{density}$$

$$l^{-3} \quad \text{concentration}$$

Centimeter (cm) to a positive degree is called conventional-spatial relations, or simply space relations, whereas cm to a negative degree is called counter-spatial, or simply counterspace relations. All the above constitute a single dimension, space. This space is bounded by a coordinate construct upon a given degree.

Hence cm to the Nth degree serves as our "space operator", operating upon a physical dimensional relation. For example, Q per cubic cm, the volume of electricity, Psi per square cm, the metrical dimension of space is applied to a physical dimension of substance. Even aether is a substance.

Mathematically,

Ql^{-3} gives Planck per cm cubed

Ψl^{-2} gives Coulomb per cm squared

In situations involving the dimension of time, the system of algebra serves well in expressing dimensional relations. It may even be said that algebra is the mathematics of time. (see Alexander McFarlane, American Association for the Advancement of Science). For situations involving the dimension of space, no suitable algebra has yet been developed.

All efforts by the great mathematicians during the 19th century were fruitless, except Oliver Heaviside's. Heaviside gave a system of vector expressions, divergence, curl, and potential, which today are WRONGLY called "Maxwell's Equations". They are not, they are Heaviside's equations, and they are NOT algebraic. But these equations have become the "Tablets of Moses", bringing from the skies the laws of electromagnetism.

But no mention is ever found on the laws of Magneto-Dielectricity, a serious drawback. (see space versor part in "Theory of Wireless Power", by E. P. Dollard). Therefore at present there is no true understanding of the spatial relationships of electricity. It is this algebraic absence that, in general, renders occult the real workings of electric induction, and specifically renders occult the work of Nikola Tesla. Space is then, the final frontier.

There is only ONE DIMENSION of Space

- Space raised to the positive exponent is simply called space (acre).
- Space raised to the negative exponent is called counter-space (per sq. centimeter).

There is only ONE DIMENSION of Time

- Time multiplied by positive one, is called forward time.
- Time multiplied by negative one is called reverse time.

Space is multiplicative (exponential) and time is additive (linear). One dimension of time(second), one dimension of space (centimeter). System of base one numbers converts the dimension into forward, backward, counter, etc.

Transmission 6

Dimensional Meanings

So far we have strongly emphasized dimensions and dimensional relations. Dimensional representation is the most direct method of analysis and synthesis with regard to the electric phenomena.

Electrical Engineering has four primary dimensions,

Metrical
    1) time, t, second
    2) space, l, centimeter

Substantial
    3) magnetism, Φ, Weber
    4) dielectricity, Ψ, Coulomb

There are no other electrical dimensions, that is it!

The electric-dimensional RELATIONS are derived from these primary dimensions, but <u>the relations are not new dimensions</u>. There are no others than "the four".

Consider a certain hypothetical transmission system, the Integratron system of George Van Tassel. (not the Goddess temple "Integratron" of today) The Integratron affects transmission around space. Let us say one is on earth and another is on Mars. If you enter the "in door" on the earth unit, you exit the "out-door" on the mars unit.

In performing this operation you did NOT travel from Earth to Mars thru any intervening dimensional relation of space. No velocity or space per time, was effected. However time has not been altered so it may be said that the dimension of space has been cancelled out. Space was the transmission obstacle and the electro-geometrical structure of the Integratron neutralized the dimension of space.

This is called a SPACE SCALAR, no variation in space.

Another example, long distance D.C. power transmission. Long distance power transmission utilizing alternating current suffers from the effects of electro-motive force, E, and the displacement current, I, both time derivatives.

The compounding of E & I over long distances results in serious transmission impairments. For D.C. the dimension of time is eliminated. Thus the dimensional relations involving time such as E and I disappear, but not into "another dimension", there is none.

D.C. has zero frequency; hence, it has no relation to time. It is eternal, invariable, and constant.

Direct Current is a TIME SCALAR, no variation in time.

Dimensional Relations of Volts and Amperes in Time and Space

We have heretofore firmly established a concept of the two fundamental laws of induction, Faraday's law of electro-magnetic induction and Maxwell's law of magneto-dielectric induction. Electro-motive force E is a magnetic reactance to a change in the net quantity of magnetic induction, displacement current I is a dielectric susceptance to a change in the net quantity of dielectric induction. E in Volts and I in Amperes are the result of inductive variation with respect to TIME.

However, also in the same units of volts and amperes exist the electrostatic potential, e in Volts, and the magneto-motive force (M.M.F.), i in Amperes. This situation does not recommend itself.

The E.M.F., E, and the M.M.F., i, have a conjugate relation thru the metallic structure of the electric system. The displacement I and the electro-static potential e have a conjugate relation within the dielectric structure of the electric system. E and i in the metal, e and I in the insulator. This suggests the electric activity E times i and another activity e times I, both in Watts. Conversely, electrical activities of E times I, as well as e times i suggest themselves. Hence we have arrived at a four polar form of electrical activity. Four distinct wattages. These represent the four terms of the "Telegraph Equation" of Oliver Heaviside in their primordial form.

The reactance E.M.F., E, via the dimension of time, T', gives rise to an electro-static potential, e, across the dimension of space, l. In conjugate form the susceptance displacement, I, via the dimension of time, T'', gives rise to a M.M.F., i, across the dimension of space, l. It may be said that E & I are the cause, where e & i are the effect, chicken or egg.

The metallic-dielectric geometric structure bounding the electric field of induction engenders the mechanical forces developed by this bounded field. Where E and I are strictly electric forces, e and i give rise to mechanical forces upon physical matter thru the dimension of SPACE. It is that i and e are the seat of magnetic forces pushing the metallic and of dielectric force pulling the metallic. i is a pushing force, e is a pulling force.

How does the dimension of space enter into e and i, both spawned of the dimension of time? From time to space, but Volts and Amperes in both. This is a problem yet to be solved, a dimensional complication. The terms *e* and *i* are not complete but are misrepresentations. They are figments of time. Forces due to *e* and *i* are more properly expressed as the dielectric gradient, Volts per centimeter and the magnetic gradient, Amperes per centimeter. *e* over *l* gives *d* the dielectric, and *i* over *l* gives *m*, the magnetic. *d* is the dielectric force, *m* is the magnetic force, of the dielectric field Psi in Coulombs and the magnetic field of Phi in Webers respectively.

Hence we have arrived at a pair of new dimensional relations,

<u>Dielectric force</u>, *d*, equals Volts, *e*, per cm, *l*
Or Weber per centimeter-second,

$$d = \frac{e}{l} = \frac{\Phi}{l \cdot t}$$

<u>Magnetic force</u>, *m*, equals Amperes, *i*, per cm, *l*.
Or Coulombs per centimeter-second,

$$m = \frac{i}{l} = \frac{\Psi}{l \cdot t}$$

A COMMON LANGUAGE FOR ELECTRICAL ENGINEERING

The dimension of time and its relation to space is evidenced by these dimensional expressions. Here a condition exists where dielectric force is derived magnetically, and magnetic induction is derived dielectrically, both thru the dimension of time. However, we are not interested in time here, we are only interested in space. It seems like we are stuck in a loop.

The way out is to utilize the new pair of dimensional relations, that is d the dielectric gradient and m the magnetic gradient, as primary dimensional relationships. Effort will be made along the way to express these relations in an alternate expression, thru the concept of inductance and capacitance.

If this pair, d and m, (both with no one's name associated with them or with any real definition) are taken as primary dimensional relations, then the electro-static potential e is derived as a secondary relation to d. Likewise the M.M.F. i is derived as a secondary relation to m.

It is then as follows,

e equals the product of d and cm,
$$e = dl$$

As well as,

e **equals Volt-centimeter per centimeter,**

$$e = \frac{e}{l} \cdot l$$

and

i **equals the product of *m* and cm,**

$$i = ml$$

As well as,

i **equals Ampere-centimeter per centimeter,**

$$i = \frac{i}{l} \cdot l$$

These dimensional relations give rise to centimeter per centimeter. What can we make of this, a space scalar. This dimensional condition represents a SPACE INTEGRAL.

Integration, as it is known, is derived from the Newton-Leibniz concepts and represents the inverse of differentiation. This application to cm per cm is called the line integral of *d* or *m*. **Integration is best avoided**.

What we are doing is this; the electric forces dimensionally are in PER CENTIMETERS, that is, in a counter-spatial form. In the integration, the product of the counterspace span in per cm is multiplied by spatial distance of the span. Per cm times cm, the product of counterspace and space, results in a dimensional cancellation, or numeric. This is called a SPACE SCALAR condition. It is dimensionless, but possesses an "angle".

In conclusion, the force exists in a counter-spatial gradient, whereas the potential, or M.M.F., exists in a spatial distance. The potential, *e*, is the integral of the force *d*, and the M.M.F., *i*, is the integral of the force *m*.

To understand this more thoroughly, read the introductory chapters of Heaviside's "E.M. Theory, volume one".

Impedance and Admittance

At this point the next level of dimensional relations can be derived from the primary dimensional relations given thus far.

(I) The law of electro-magnetic induction, Faraday's law, that is the electro-motive force E, in volts, is given by the proportionality (ratio) of the total quantity of magnetic induction Phi, to the time rate of the gain or loss of this quantity of magnetic induction, in per second. The voltage E is given by the rate of variation of magnetism.

Change in Magnetism is Volts of E.M.F.,

$$E = \frac{\Phi}{t} \quad \text{Volts}$$

(II) The law of magneto-dielectric induction, Maxwell's law. That is, the displacement current I, in amperes, is given by the proportionality (ratio) of the total quantity of dielectric induction Psi, to the time rate of the gain or loss of this quantity of dielectric induction, in per second. The current, I, is given by the time rate of variation of Dielectricity.

Change in Dielectricity is Amperes of displacement,

$$I = \frac{\Psi}{t} \qquad \text{Amperes}$$

In both cases "quickness" is the factor of direct proportionality. Example, 120 volts at 60 cycles per second applied to a transformer winding results in a greater rate of change in magnetism than 110 volts, 60 cycles applied to the same winding, despite both being 60 cycles. Why? The slope of 120VAC is greater than 110VAC. Try it on your oscilloscope and see.

E and I are not to be considered opposites of each other, but they exist in a COMPLEMENTARY-SYMMETRY form. The four pole archetype of electricity shows itself in that there is E and e or I and i. This leads to the answer for our second question, the null force condition, that is what ratio of E to I, and thus e to i give rise to a cancellation of "e pulls" and "i pushes". Another ratio to be investigated. Taking the ratio of the E.M.F. E, and the displacement I, that is E over I, we have evoked "Ohms Law": The dimensional relation of E.M.F., Phi over T; divided by the dimensional relation of displacement, Psi over T.

This results in a new dimensional relation. This relation is known as the IMPEDANCE, Z, in OHMS.

E per *I* equals *Z*

For a given product of *E* and *I* in Watts, we may have a large *E* and a small *I*, a high impedance, or we may have a small *E* and a large *I*, a low impedance. Hence, a unit of power (activity) in Watts may be in the form of a high impedance (12KV, 1 Amp) or a low impedance (1KV, 12 Amp), both the same power (12 KVA).

Think of the transmission in your car. The engine is delivering 20 horsepower (activity) and this is delivered to the wheels. The engine is running 1800 R.P.M. (volts), but the drive shaft is running 180 R.P.M. (volts). The engine is a high impedance, the driveshaft a low impedance, but the power is 20 HP in both.

We call this an IMPEDANCE TRANSFORMATION and this is effected by what is known as a TRANSFORMER, (the transmission, it has a RATIO of ten to one).

The dimensional relation of impedance, Z in Ohms, can be expressed in an alternate manner from the primary dimensions. *E* divided by *I* equals Z,

Ohms Law. But we have dimensionally, that the E.M.F., E, in volts is given by Faraday's Law.

That is,

$$E = \frac{\Phi}{t} \quad \text{Volts}$$

Likewise the displacement current, I, in Amperes is given by Maxwell's Law,

$$I = \frac{\Psi}{t} \quad \text{Amperes}$$

Taking the ratio of E over I and substituting, the impedance is given by,

Weber-second per Coulomb-second equals Ohms,

$$\frac{E}{I} = \frac{\Phi t}{\Psi t} = Z \quad \text{Ohms}$$

Here, dimensionally speaking, we have second per second which is thus dimensionless, or scalar, a

TIME SCALAR. Hence the primary dimensional expression for impedance, Z in Ohms, is given as,

Webers per Coulomb equals Ohms,

$$\frac{\Phi}{\Psi} = Z \quad \text{Ohms}$$

Hereby the impedance of the electric field of induction is defined as the ratio of the total magnetic induction to the total dielectric induction, Phi over Psi gives Ohms Z. This is known as the characteristic impedance of the electric field of induction.

[It must be remembered that the scalar term of seconds per second expresses the hysteresis angle between the time frame for E and the time frame for I, as the pair weave their dance thru the dimension of time (note, get that 2D or 3D out of your head, we are in the dimension of time!) The ratio Z, the impedance, is therefore a "directed quantity" in the dimension of time. This is to say the impedance has magnitude and a position in time. Listen to Bach organ music for further as this is too complex for now.]

Since arriving at the concept of impedance it may be asked. what results from its inverse I over E, the ratio of displacement current to electromotive force? This ratio is called the ADMITTANCE, Y, of the electrical system. Following the same path dimensionally as was done with impedance it is,

Coulomb-second Per Weber-second equals Siemens,

$$\frac{\Psi t}{\Phi t} = Y \qquad \text{Siemens}$$

Hence, the admittance Y is given dimensionally by,

Coulombs per Weber equals Siemens,

$$\frac{\Psi}{\Phi} = Y \qquad \text{Siemens}$$

Admittance Y in Siemens is the ratio of Psi to Phi, the ratio of the dielectric field to the magnetic field of the electric field of induction. This is called the characteristic admittance of the electric field.

As for the scalar term of seconds per second, the same situation exists as with the impedance, Z in Ohms. It is however that there is also a time "angle" between the time frame of impedance, Z, in Ohms and admittance, Y, in Siemens, just as there is with

Volts and Amperes. Hereby results that the impedance is NOT just the inverse of the admittance, that is, Z is NOT one over Y, they are MIRROR IMAGES. Look in the mirror. Your head is up, your feet are down, but your right is left and your left is right. This is much too complex.

In conclusion, the impedance and admittance serve as proportionality factors between the magnetic and the dielectric, or the dielectric and the magnetic, fields respectively.

The End of Time

This dimension of time, and its dimensional relation to electric phenomena, thus far has been the primary discussion. This metrical dimension, in seconds, and its relation to the substantial dimensions of Q, Psi and Phi has yielded a set of fundamental electrical relationships. As of yet the dimension of space is regarded as a frontier, occult from the human mind.

The idea of the dimension of time is a relatively modern development, beginning with the "age of clocks" in Europe. Out of this evolved the epic time dimensional works of J.S. Bach, 1685 to 1750. Following Bach, the Newton-Leibniz system of mathematics allowed for an analytical treatment of the dimension of time. The development of a base one number system, such as given by A. MacFarlane, and a system of electrical dimensional relations by O. Heaviside, led to the epic time dimensional works of C.P. Steinmetz 1865 to 1923. Steinmetz writing in "Theory and Calculation of Transient Electric Phenomena" stands as an algebraic analog to Bach's last work, "Die Kunst der Fuge", BWV – 1080. Time in both forward and reverse, as well as multiple hysteretic time frames, within a complimentary-symmetrical geometric construct are portrayed in both the works of J.S. Bach and C.P. Steinmetz.

[A persistent rumor exists, starting with Edison's attempts to make contact with the dead. An unknown Christian organization contracted Steinmetz to develop an apparatus to allow viewing in present time the life of Jesus Christ in his time. Quite a story but no possible verification. However, in the realm of fiction, it was noticed that an old 1963 Outer Limits episode titled "Borderland" portrayed a layout like Steinmetz may have used in his G.E. laboratory. It is unfortunate that few, if any, have, or will follow the footsteps of the giants, they are too deep. We will continue to follow their path, hard as it may be.]

Up to this point in time these time sequential internet transmissions have been intent in presenting, in a hypo-complex fashion, the primary electrical relations. The only unique dimensional relation has been the Planck, Q. The Planck is defined as the undivided quantity, Q, of the total electric induction. In a Biblical sense an analog is, "and in the beginning…" Every other substantial relation found its derivation in the Planck. A pair, Phi and Psi, was arrived at by a divorce, four more by direct interaction of a substantial dimension with the metrical dimension of time, these being E, I, W, P, and finally two more, Z and Y, were derived in a pair of paths, one path thru the metrical dimension of time, the other path thru direct relations of the primary substantial dimensions of Phi and Psi. These give us the ONE and a following set of EIGHT.

It is important to remember not to become enamored to our models, or the models of others, with regard to the nature of the inductive process, or the nature of the medium which engendered it. These constructs, here and elsewhere, are illustrative, not concrete, but that is not to say that concrete constructs should not be sought. The primary intent and effort here is the establishment of concrete engineering formulation applied to the development of specific electrical apparatus. It is goal orientated much like the actions of large aggressive reptiles.

Concluding this sequence of time dimensional relationships, this summary and table are given.

Summary Statements and Mathematical Expressions

The One: The Undivided Quantity of Electric Induction,

Given as,

> The Planck, Q, in Weber-Coulomb

$$Q = \Phi\Psi \quad \text{Planck}$$

The Primary Pair of Dimensions,

> *One: The Total Dielectric Induction*

Given as,

> The Coulomb, Psi, Ψ, a primary dimension

> *Two: The Total Magnetic Induction*

Given as,

> The Weber, Phi, Φ, a primary dimension

The First Set of Secondary Dimensional Relations,

One: The displacement current

Given as,

<u>The Ampere</u>, I, in Coulomb per second

$$I = \frac{\Psi}{t} \quad \text{Amperes}$$

Two: The electromotive force

Given as,

<u>The Volt</u>, E, in Weber per second

$$E = \frac{\Phi}{t} \quad \text{Volts}$$

The Second Set of Secondary Dimensional Relations,

Three: The energy

Given as,

A COMMON LANGUAGE FOR ELECTRICAL ENGINEERING

<u>The Joule</u>, W, in Planck per second

$$W = \frac{Q}{t} \quad \text{Joule}$$

Four: The activity (power)

Given as,

<u>The Watt</u>, P, in Joule per second

$$P = \frac{W}{t} \quad \text{Watts}$$

And Finally, the Pair with Dual Derivation,

One: The impedance

Given as,

<u>The Ohm</u>, Z, in Volt per Ampere, or in, Weber per Coulomb,

$$Z = \frac{E}{I} = \frac{\Phi}{\Psi} \quad \text{Ohm}$$

Two: The admittance

Given as,

The Siemens, Y, in Ampere per Volt, or in, Coulomb per Weber

$$Y = \frac{I}{E} = \frac{\Psi}{\Phi} \quad \text{Siemens}$$

Also, throughout, **the dimensional relation for time** is given as,

The complex frequency in Neper-Radians per second, v,

And thus closes our system of electrical units in a dimensional relation with time.

Table 1

Dielectric	Electric	Magnetic
	Planck, Q	
Coulomb, Ψ		Weber, Φ
Ampere, I		Volt, E
Siemens, Y		Ohm, Z
	Joule, W	
	Watt, P	

(Neper-Radians per Second)

Space

Our discussion on the theory of electric phenomena will now shift from the metrical dimension of time and develop the application of the metrical dimension of space. As a frontier, space will be treated in an experimental manner. One is cautioned against being, or becoming, enamored to any particular concept of space presented here or elsewhere.

The importance of the study of J.J. Thomson, "Electricity and Matter", and O. Heaviside, "E-M Theory", volume 1, (introductory chapters), cannot be overemphasized. Do your homework. And as said by Heaviside, it is said here; "If you think these are hard to read, well, they are even harder to write" (in the front seat of my car).

Greater complexity is encountered in the application of the dimension of space than that of time. Space relations are not well understood in the contemporary engineering world, such as curved space and 2D, 3D thinking has all but eliminated any real comprehension of the application of the dimension of space to electrical study.

Relations in the dimension of time are ADDITIVE, relations of forward and reverse. Time relations are linear, so to speak, along a versor path in time. Not so in the dimension of space, here it is MULTIPLICATIVE. Time relations are arithmetic progressions, whereas space relations are geometric progressions.

Consider a numerical illustration,

> For the time relations, it is:
> 3 and 3 gives 6

> For space relations, it is:
> 3 and 3 gives 9

Two distinctively different results,

> In time, it is expressed as:
> 3 plus 3 equals 6

> In space it is expressed as:
> 3 times 3 equals 9

Taking an alternate approach, we have for time, a pair of 3's, that is, two 3's.

The same for space, two 3's,

> Then it is for time:
> 3 times 2 equals 6

> And for space it is:
> 3 to the 2nd power equals 9

Hence, it can be stated that the time relations are LINEAR, whereas the space relations are LOGARITHMIC.

In working with time dimension relations there has been developed a versor algebra, an "algebra of time". The development of this time algebra is primarily due to the efforts of Charles Proteus

A COMMON LANGUAGE FOR ELECTRICAL ENGINEERING

Steinmetz of General Electric Corp. (also see A. MacFarlane, "The Imaginary of Algebra").

Versor algebra utilizes a BASE ONE number system, that is, the number ONE raised to a specific power. In common use is the base TWO number system, binary, and the base TEN number system, decimal. But a base one number system seems unlikely, it can only give one. This is exactly what we want in a versor system, as will be shown.

The versor concept applied to time relations has been established in engineering work for practical use, but not so for space relations. The concept of space versor algebra jammed up and seized with what is known as QUATERNIONS. Development of space versors "finished" with quaternions (see A. MacFarlane "Principals of the Algebra of Physics") and little else progressed up to the "Theory of Wireless Power" by E.P. Dollard where a space versor algebra is presented. Note that versors are NOT VECTORS.

So as to arrive at an understanding of versor algebra in general it is instructive to consider known concepts of versor algebra for time. Consider a second order time relation. This is given as, $\sqrt{+1}$ a seemingly useless statement. An alternate expression is given as, ONE raised to the one half power, let us say raised to the PER TWO power, $1^{\frac{1}{2}}$.

A COMMON LANGUAGE FOR ELECTRICAL ENGINEERING

This is a second order relationship and by definition must give a pair of roots, that is, two distinct values, thus the roots are given as (1) Positive one, and (2) Negative one. We will call negative one an IMAGINARY NUMBER. Both roots, (1) and (2) are more properly named VERSOR OPERATIONS. Their magnitude is unity and they are dimensionless. Therefore, by application to the concept of the additive process, it has now been extended to include subtraction. (plus and minus). We have arrived at an expression for time phase opposition.

Extending these ideas results in a versor expression for fourth order time relations, the fourth root of positive one. This is ONE to the ONE FOURTH power (one to the PER FOUR power), $1^{\frac{1}{4}}$. This can be partially resolved into a pair of square roots, each of which have two roots, giving the required four roots.

It is given that the square root of positive one produces a pair of roots, that is,

1) + 1
2) - 1

Now, in addition, is the second square root, the square root of negative one, this producing a pair of roots,

1) + j
2) - j

Hereby we arrive at the required four roots produced by the fourth root of positive one, four to the one fourth.

These are given as,

0) + 1
1) + j
2) - 1
3) - j

Here the concept of the additive process, addition, is now further generalized. There is more than only plus and minus. This versor expression is for time phase quadrature and for time phase opposition. We now see versors as a degree of lead or lag on a given order of time, here in fourth order time. One to the one fourth power gives the four quadrature versor operators. CAUTION: do not view this as 2D or 4D time, this is NOT a VECTORIAL system. Further on this is given in the "Symbolic Representation" papers by E.P. Dollard.

The versor algebra here given thus far has been orientated to represent the dimension of time. Versor algebra is based upon degrees or orders of the dimension to which they are applied. The following orders of space, in terms of perceptual understanding begins the degrees or divisions of the dimension of space. The fundamental unit of space here will be the centimeter, or cm.

I – First order of space:

(a) DISTANCE Space to the positive first power, l^1.

(b) SPAN Space to the negative first power, l^{-1}.

II – Second order space:

(a) AREA Space to the positive second power, l^2.

(b) DENSITY Space to the negative second power, l^{-1}.

III – Third order space:

(a) VOLUME Space to the positive third power, l^3.

(b) CONCENTRATION Space to the negative third power, l^{-3}.

IV – Fourth order space:

(a) Space to the positive fourth power, l^4.

(b) Space to the negative fourth power, l^{-4}.

Finally Nth order space.

The (a) group are the spatial relations, and the (b) Group are the counter-spatial relations. Higher order space, such as fourth order lack definition, it is undefined space. Einstein 4D space, crystal structured? It remains undefined!

These orders of space can be worked into a simple tale:

> I ran out of orange juice, bummer. Now I got to drive a DISTANCE of 15 MILES to the lizard pit (store). Let's take the poodle, along for a chance to get out of the house. Along the way I'll count the number of poles on that old J carrier line, they have a SPAN of one

PER 150 FEET. The poodle has got to stay in the car, that 1000 ACRE AREA of land behind the store is infested with coyotes. They have a population DENSITY of one PER 50 ACRE. This store only sells juice in a VOLUME of half GALLONS. The CONCENTRATION of vitamin C in this juice is one unit daily requirement PER 8 CUBIC OUNCE serving. When considering next the fourth order of space, the dog begins to bark his head off at some UNKNOWN THING. Stupid poodle, let's go.

Break…..

…..more follows on Space.

Transmission 7

In Space, Heaviside

The Lord God saith to Moses: Thus rose the beast, and it arose as a giant one winged parrot, this hewn of solid plutonium. She delivereth the beast as an idol of worship to her supreme power. The surrounding multitudes casteth off their garments so as to feel the warmth of its radiant tongues of fire. It came to pass that the beast smote the multitudes by burning their eyes and leaving them to wander blind for eternity.

Verily this is the present situation for the science of electricity, just as told in the parable. Oliver Heaviside, a spurned English theoretician, devoted considerable effort toward straightening the path of progress in electrical science. He endeavored to lead the blind out of darkness, but their restored sight was only black and white. The idea of electricity as a flow of electrons in a conductor was regarded by Heaviside as a psychosis. This encouraged Heaviside to begin a series of writings to "de-program the Moonies". This often utilized "Tragic Tales" to make his idea clear about mathematical futility. "One such tale" is of a young child, "and this child would not smile. So his father beat him with a strap, in an effort

A COMMON LANGUAGE FOR ELECTRICAL ENGINEERING

to make him smile. But the child would not smile. Later he was eaten by a lion". This tale is what Heaviside thought about the mathematical system of QUATERNIONS.

Quaternions represent a primordial form of a versor space algebra. It is however ultimately vectoral. The system of quaternions was a retrofit of the existing concepts of versor algebra that originally were adaptable to time, this to provide analogous versor algebra useful for space. It is really not much use for this purpose. Oliver Heaviside despised quaternions, leading him to develop his own space mathematics. The work of J.C. Maxwell was encumbered by his reliance upon the use of the quaternion system. It is this complication with Maxwell that initiated Heaviside's efforts to "remove the baggage" that encumbered the understanding of Maxwell. Heaviside produced his "vector system" for this effort. He received little credit for his important work, and his famous set of equations are wrongly known as "Maxwell's Equations", ask any parrot.

Heaviside's vector system sought to provide a lucid description to the Maxwellian concept of electro-magnetism, or rather the Heaviside concept of the Maxwellian concept of electro-magnetism. That is transverse electro-magnetism. Plancks, or cross product relations. But this vector system has no application to longitudinal magneto-dielectricity. But is not this the realm that gave Maxwell his fame? Displacement current, current in empty space, is the Maxwellian concept that leads to longitudinal waves. It is most likely that most of what is thought to be understood about Maxwell's work is in reality the ideas of others. Absolutely no mention should be made of anything Maxwellian without directly quoting Maxwell himself through his writings! To study his work in its entirety will take a lifetime. So mute thyself.

    The dis-informers, Soviet Scalar Xenophobes (SSX), heap criticism upon Oliver Heaviside for his one sided representation of the works of J.C. Maxwell. Their real motive is to take one off the "Heaviside trail", but the coyote wants to run that path. Obviously this series of transmissions via internet is in the full "attitude" of Oliver Heaviside. Forgotten in prehistory is the effort of 19th century mathematicians given to space math. Major figures were Grassman, Hamilton, Tait, and MacFarlane. Review of their works provides a better understanding of what has been inherited. Only what Maxwell utilized out of necessity, and Heaviside developed out of deliberation exists today. Longitudinal electric waves are considered non-existent, a complete impossibility. But that is what propagates between the plates of any condenser or between the windings of any transformer, physical realities.

So today we labor under an absolutely one sided view of electricity. It cannot stand upright just as a bird with only one wing cannot fly. The entire misconception has now matured, become "frozen in stone" as an idol of worship to the supreme power of the pedant. In reality it is a monolith of self-edification, taught in every university.

The pedantic assault upon the pioneering efforts of Heaviside, Tesla, and Steinmetz is worthy of closer examination. Two particular characters, supreme pedants sink to the "bottom of the bowl". One is William Preece of the British Royal Society. Preece championed the entirely lopsided view prevalent in the "misunderstood, untested, and infested with bugs" undersea cable telegraph technology. This errant concept nearly ruined the trans-oceanic telegraph industry. It took the work of Oliver Heaviside to "balance the equation" thereby allowing a working telegraphic system. He did this through his serial writings known as "Electro-Magnetic Induction and its Propagation". These led to his most important development, the "Telegraph Equation". William Preece F.R.S. censured the work of Oliver Heaviside and went so far as to attempt to make his own concept a law.

America's most (wanted) noteworthy pedant is Michael Pupin, of Columbia University. Pupin was a supreme pedant. He repeatedly assailed Tesla with rude attacks. He even went so far as to perform the same upon C.P. Steinmetz. Steinmetz received harsh criticism for his theory of hysteresis, which not only gave Steinmetz his world fame, but also saved the infant electrical industry from ruin. Pupin declared the important transformer equations developed by Steinmetz as un-Maxwell, heretical. But it is these ideas that lay the foundations of electrical engineering. Pupin's crowning swine behavior is his treatment of Heaviside. Pupin takes Heaviside's telegraph equation, repackages it as a transmission concept, patents its implementation, and then sells it to American Telephone & Telegraph for $25,000. Long distance telephone is born, with not one word of Heaviside. In final disgust, during the "Einstein Age", Heaviside removed his furniture from the house and sat on granite blocks, then painted his fingernails pink. Later he was stoned to death by a pack of youngsters.

Application of Space to the Electric Dimensional Relations

In order to gain an understanding of the electric field of induction, a concept of the distribution of this induction in the dimension of space must be developed. An example is a 200 mile long power line. It has a span of 600 feet between towers. This is a 230 kilovolt, 60 cycle/sec, 3 phase line. It can be shown that for each span of line between supporting structures there exist an electro-motive force, *E*, in Volts, this in series along this span, and a displacement current, *I*, in Amperes, this in shunt along this span. The series E.M.F., and the shunt displacement of each span compound with each successive span. The total E.M.F. and total displacement for the entire length, 200 miles, of the line is found by integrating over the total number of spans. However, this integrated value is not given by the simple addition of the individual E.M.F.s and displacements developed by each of the individual spans.

Here we find an exponential function of space determines the relation between the individual values, and the total values of E.M.F. and displacement. These considerations are developed by Charles P. Steinmetz in his "Theory and Calculation of Transient Electric Phenomena" book, in particular the chapter on "Transients in Space".

The general problem of the representation in space is given by the introductory part of "Transients in Space" and also by Ernst Guillemin in the introductory chapters in his "Communications

Networks" vol. 2. Read these, they are a most important study. These writings form the basis for the theories of electrical engineering utilized today.

The metrical dimension of space is most often considered as a VOLUME, this representing an enclosed quantity of space. This space is filled with something substantive, often which must be paid for, such as a gallon of milk. The milk is the substantial dimension, the "throw away" gallon container is the metrical dimension. In general, this volume of space is considered a cubic quantity, or boundary, this such as a cubic foot, cubic yard, cubic centimeter, and etc. It is habitual to express a volume in cubic terms, this in three mutually perpendicular coordinates, wrongly called "dimensions". It is also habitual to express electric relations in the same manner, a corner of a cube, such as Psi, Phi, and Q. This now is three dimensional, a 3D relation, since Psi, Phi, and Q are dimensions.

The cubic relation itself has no substantive dimensions, it is only the metrical dimension of space expressed by a group of three mutually perpendicular coordinates. This is important.

It is given here that a one centimeter cube is the elemental unit of the dimension of space, a volume of one cubic cm. This is about the size of a common sugar cube, but instead of sugar, this cube is filled with electric induction. It is a cube of electricity.

What exists outside the boundaries of this cube is for now unknown, it is excluded by the boundaries. For most of the examples that follow, all space is filled with 10-C transformer oil, the dielectric. All boundaries enclosing, or dividing, this space are sheet copper, the metallic. Here given is the metallic-dielectric geometry, such as a power transformer, or a static condenser, two fundamental apparatus in electrical work.

In the case of the 200 mile long A.C. power line the basic element of space is the span. This is first order space. Here it is given as per 600 feet, this now a unit value. It now equals one, one span. This unit value is known as a differential element, it is indivisible, the smallest "line on the ruler". It relates to the Newton-Leibniz concept of the infinitesimals.

It is considered that cubic, or third order space, is the most general expression of space, a metrical dimension. Since the ordinary transformers and condensers utilized in power engineering are of considerable volume, it is then allowable to consider one cubic cm. as a differential element, that is, an infinitesimal quantity of space. Taking the, one cubic cm of space, as a unit value, gives the differential element of the metrical dimension of space. It is hereby about the size of a sugar cube, but filled with 10-C oil. Hence the VOLUME is given as ONE cubic cm, the AREA as ONE square cm, the DISTANCE as

ONE cm, the SPAN as ONE per cm, the DENSITY as ONE per square cm, and the CONCENTRATION as ONE per cubic cm, all faces, corners, spacings, and etc. of this unit cube are ONE. Hence our differential, indivisible, element of the metrical dimension of space. All orders, or powers, of this space equals one, one squared is one, one cubed is one, and etc. All are one.

    This may just as well have been a cubic yard, or a cubic nanometer. The consideration of "unit value" is to reduce the size to the point to which there is no distinguishable variation of the substantive dimension with respect to the unit of the metrical dimension of space.

    It is then a space scalar condition, no variation in space. For example, consider the 200 mile long A.C. power line. It has a propagation velocity very near that of light. For a frequency of 60 cycles per second, this gives the wavelength as 2880 miles in length. The total distance of this line is 200 miles, this a significant fraction of a quarter wave or an impedance to admittance transformation. However the per 600 feet of a span is an infinitesimal fraction of the quarter wave distance. Hence the distance between towers, the spans, serve as the differential element. It is then 600 feet is of unit value, indivisible. There are no intervening towers.

A COMMON LANGUAGE FOR ELECTRICAL ENGINEERING

No perceptible variation of the series E.M.F., E in Volts, or the shunt displacement, I in Amperes, exist along this 600 foot span of A.C. power line. Hereby it is said the E.M.F. per span, or the displacement per span. In the general case it is given as Volts per span and Amperes per span. These dimensional relations represent the voltage gradient and current gradient along the length of line.

Dimensionally it is,

1) Volts per cm, $\quad\dfrac{E}{l}$

and

2) Amperes per cm, $\quad\dfrac{I}{l}$

It should be noted that this pair of gradients exist in space quadrature to the previous given gradients.

The dielectric gradient is,

$$d = \frac{e}{l}$$

And the magnetic gradient is,

$$m = \frac{i}{l}$$

This is a fundamental relation in electro-magnetic induction and its propagation.

It can be seen that each span has a back E.M.F. in series with the power flow, and a displacement, or charging, current in shunt with the power flow. These are a consequence of the electric field of induction in a time rate of variation, the 60 cycle, or 377 radians per second.

<u>These are transmission impairments</u> and give rise to a delay in propagation which progressively compounds down the line, from span to span. These differential elements, or spans, must be summed up, or INTEGRATED, in order to determine the total E.M.F., total displacement, and the total delay in propagation. This is not so easy of a task. Now for higher orders of space the situation is that order more difficult.

In the application of the metrical dimension of space to the substantial dimensions of electricity, the concept of magnetic inductance, and electro-static capacity, are utilized. Steinmetz, in his "Impulses, Waves, and Discharges", established the inductance and the capacitance as the "Energy Storage" coefficients of the electric field of induction. It must be noted that here the term ELECTRIC FIELD is NOT the "electro-static" field, it is the union of the dielectric and magnetic fields of induction. Erase the "electric field" wording of the one wing parrot. These two distinct dimensional relations, the INDUCTANCE and the CAPACITANCE serve to define the ability of bounded space to contain the electric field of induction, this field representing STORED ENERGY.

What follows here is the development of the properties related to the dielectric and magnetic fields and the interaction of these with the bounding metallic-dielectric geometry. Considerations involving energy will be arrived at later on. In this view inductance and capacitance now represent GEOMETRIC COEFFICIENTS, expressing the relation of the BOUNDING GEOMETRY with the fields of induction which it bounds. In essence inductance and capacitance are of a scalar form. Here enters the concepts of what is known as "Radionics". The inductance and capacitance each exist in distinction to the electricity itself. The inductance and capacitance ultimately serve as geometric expressions. This is important. Hereby they can be expressed as completely metrical dimensional relations, that is, having no substantive dimension.

We have of yet actually given the dimensional relations which make up inductance and capacitance. Further considerations involving the electric field have yet to be understood.

Basic Electrical Relations in Space

Let a one centimeter cube of the metrical dimension of space be given as unit space. The given metrical boundaries define a unit cube, a sugar cube with no sugar, no 10-C oil, not even any aether. This cube of space is empty, void. This unit of space, a metrical unit, is NOT in connection with any substantial dimension. This is a unit cube of void space.

It is postulated that electric induction, as a property of the aether, cannot be established without the presence of this aether, or another dielectric medium. This is to say, void space is not capable of supporting electric induction. Hereby it is reasoned that the aether is a substantive dimension, or it can be expressed as a substantive dimensional relation. The dimensional expression of the aether it is best given as a primary dimension, expressed only in a dimensional relation derived from a primary substantive dimension and its relationship to a primary metrical dimension, time or space. This is served by the concept of the Planck, Q, this in conjunction with a space versor system expressing the Planck in real (electro-magnetic) components. See E.P. Dollard "Theory of Wireless Power".

In general we have been speaking in terms of line of force that is lines of magnetic and lines of dielectric induction. It is as of yet known "how big" these lines of force may be. These lines of induction can be expressed analogously, such as a bag of uncooked spaghetti. Its individual strands serve as analogs to the individual lines of induction. Hence long, thin, strands with axes in a broadside bundle. If then the bag of spaghetti is snapped in two, when it is viewed endwise, a circular cluster of small end-sections of each strand are seen. These are elements of the strands of spaghetti.

This package of spaghetti is an analog to the total dielectric induction. Now there are 100 strands of spaghetti in this package, or boundary condition. Here Psi is then given as 100 strands, the total quantity of spaghetti. Viewing this package endwise, 100 end sections of the individual strands are seen in a circular bundle. The cross sectional area of this bundle, bounded by the package is a one square inch,

AREA. This is analogous to a dielectric flux DENSITY, the density of dielectric induction in PER SQUARE INCH. Hence the density of spaghetti is given as 100 strands per square inch. The total undivided quantity of spaghetti is 100 strands. The substantive dimension of spaghetti is in containment by the "throw away" package, this as the metrical dimension of space. Counter-spatial representation is given as 100 strands per square inch. The grocery store regards this as one indivisible unit quantity in space of spaghettic induction.

Here the substantial dimension of spaghetti, Psi, is operated upon mathematically by the metrical dimension of space, per square inch. Here arrived at is the dimensional representation of spaghetti in space, strands per square inch. This is a second order space relation, in a counter-spatial form.

A pair of dimensional relations follow from the spaghetti analogy,

One is the Dielectric Flux Density, Psi over A,

(1) Coulombs per square cm, $\dfrac{\Psi}{l^2}$

Two is the Magnetic Flux Density, Phi over A,

A COMMON LANGUAGE FOR ELECTRICAL ENGINEERING

(2) Webers per square cm, $\qquad \dfrac{\Phi}{l^2}$

Here A is the area in square cm, a second order expression of space. A primary dimensional expression for second order space is the ACRE.

Since the total electric induction, Q, is the product of the dielectricity, Psi, and of the magnetism, Phi, it may be asked, what of the product of the dielectric flux density, and the magnetic flux density?

This is given as,

(3) Coulomb per square cm, $\qquad \dfrac{\Psi}{l^2}$

<u>times</u>

Weber per square cm, $\qquad \dfrac{\Phi}{l^2}$

Substituting the dimensional relation,

(4) Coulomb-Weber, or Planck, $\qquad \Psi\Phi = Q$

Hereby gives the dimensional relation of the product of the flux densities as,

(5) Planck per quartic cm, $\dfrac{\Psi\Phi}{l^4}$

So, the product of the pair of flux densities, (1) and (2) gives rise to a fourth order space relation (5), in its counter-spatial form.

Now, what set off the poodle this time?

Transmission 8

Experimental Cubic Volumes

It has been heretofore established an existence of a volumetric, or cubical, unit of space, the metrical dimension. This unit of the dimension of space is defined as one cubic centimeter, the size of a common sugar cube. It is given this metrical cubic volume of space is void of any substantive dimension, no sugar, no 10-C oil, and even no aether, since aether is considered a substantive dimensional relation. Hence, indivisible, void, and a pure metrical unit, this is our cubic volume of EMPTY SPACE.

Consider the axiom that a field of electric induction cannot exist in the absence of the aether. Then just how does this cubic void space interact with an electric field? Since the laws of lines of force as established by Michael Faraday, and developed by J.J. Thomson, and as further established by C.P. Steinmetz, maintain that no line of force can just end in space. The lines of magnetic induction exist as closed loops, no beginning, no end, continuous expansive or contractive loops. Magnetism is a circumferential force. In a conjugate manner the lines of dielectric induction terminate upon physical surfaces, where they bond into the intra-molecular dimensions. Dielectricity is a radial force.

The Maxwell concept of electro-magnetic induction and its propagation gives an altered concept of the nature of dielectric induction. In this situation the lines of dielectric induction may also terminate upon themselves, forming closed curves in a manner analogous to the loops of magnetic induction. This condition is a necessity for the propagation of electro-magnetic waves in a dielectric medium, (sugar, oil, aether, etc.) without guiding metallic structures (wires, waveguides, etc.) This eliminates the "charge carrier", that is, the dielectric induction is now completely independent of any terminal surfaces.

The dielectric induction is now completely dielectric. This is the fundamental concept underlying the Maxwell theory of electromagnetism. It is here that J.C. Maxwell found his fame. But the pedant tells us just the exact opposite! So intent is this mind-state in forcing a "materialism" upon electrical theory that Maxwell's work is reworked to suit this view.

It may be logically inferred hereby that, for the condition of a cubic volume of space, the lines of magnetic, and the lines of electric, induction must bend around this cubic void of space. These lines cannot be interrupted or broken by this void. Hence by the insertion of a cubic void into a space supporting electric induction the lines of force are pushed aside. The overall induction in the supportive space is then hereby reduced, since now there is a unit volume less of this space. This is to say that the inductivity of the supportive space is reduced by the insertion of an aetherless cubic volume of space, <u>the cube of empty space</u>.

    Consider certain experimental configurations. One configuration consists of a widely spaced pair of laser produced beams of light, side by side traveling through the aetheric medium. The second configuration is a pair of one square centimeter copper plates. These two plates face each other squarely and are separated by a span of one centimeter. This defines a partial boundary for our one cubic centimeter, or unit, cube. Hence any unit cube volume can be inserted between the pair of one square cm copper plates. It is also given that all space within and surrounding these copper squares is void, no sugar, no oil, no aether, just empty space.

    In our first experimental configuration we have a set of three unit cubes, one is filled with 10-C oil, the second is filled with aether, and the third is void. Taking the side by side spaced laser beams, we measure the speed, or time delay of propagation of both beams through supporting aether through which they propagate. Here, both arrive at the end point at the same time, thus propagating at identical velocities of propagation. First, take the unit cube of oil and insert it into beam number one, leaving beam number two unchanged. It is hereby found that beam one arrives delayed in time relative to the arrival time of beam two. Here it can be inferred that light travels slower in the oil. By measurement it is found to be about 70 percent of the light velocity in the aether.

Next, take a unit cube of aether and insert this cube into beam one, again leaving beam two unaltered. Obviously both beams arrive at the same time since both propagate through only aether.

Finally, take a unit cube of void space and insert it into beam one, beam two again unaltered. The poodle begins to bark. We now have two distinctly opposing possible outcomes.

(A) Beam one is stopped at the facing boundary of the cubic void. No beam one is detected at the receiving end. Now it may be asked, what became of beam one? Was it sent back, or was it consumed, thus in violation of the Law of Energy Perpetuity? This we are unable to answer.

(B) Beam one arrives advanced in time relative to the arrival time of beam two, this to say, that the propagation through the void space is now instantaneous, in other words with an infinite (undefined) velocity. It takes no time to span the distance of the unit void space. How is this possible?

Now we take our next experimental configuration, the pair of parallel one square cm. copper plates, these in void space. Thus far we have no concept defining capacitance, but we do possess a capacitance meter. How fortunate! Upon connecting this instrument to the unit copper plates in a void it is found that this metallic-dielectric configuration has zero capacity. This is understandable since we now have no dielectric, and hence, no dielectric induction.

Next, we insert a unit cube of aether between the unit square copper plates. Now the instrument indicates one electro-static unit of capacitance, this as expected.

Finally, we insert a unit cube of 10-C oil between the unit square copper plates. Now the instrument indicates an increase in capacitance over that of the aether. This increase in capacitance is in EXACT proportion to the square (second power) of the decrease in the velocity of light through the same identical cube of oil. It is then given, the change in the velocity of light through a dielectric medium is the square root of the inverse of the change in capacitance affected by this dielectric medium. Hence capacitance exists in an inverse relationship with the velocity of light in a given medium. Zero capacitance, infinite velocity.

Hereby this dimensional relation is given as,

Seconds Squared Per Centimeters Squared $\quad \dfrac{t^2}{l^2}$

This is to say, one over the speed of light squared, that is, one over c squared. Here it is useful to take the speed of light as a unit value, or one. See for example, C.P. Steinmetz's "Impulses, Waves, and Discharges", chapter on "Velocity Measure". It is in this relationship between luminal velocity and electro-static capacity that we find the luminal velocity concepts of the relativists, the c squared in the E equals mc squared. Call it a "dimensional fluke" if you wish. However, capacitance is forever married to the velocity of light, to one over c squared.

Investigating dielectric capacitance a bit further, consider an experiment of Ben Franklin, the father of the electro-static condenser. Here we will dispel the "electronics nerd" concept that a capacitor stores "electrons" in its plates. Taking the pair of copper plates as in the previous experiment, but now we have two pairs of plates, one pair of plates distant from the other pair of plates. Upon one pair of plates is imposed an electro-static potential between them. The cube of 10-C oil is inserted between this "charged" set of plates. This hereby establishes a dielectric field of induction within the unit cube of 10-C oil. Now we then remove this cube of oil, withdrawing it from the space bounded by the charged pair of copper plates, and taking this unit cube of oil, it is then inserted into the space bounded by the other uncharged pair of plates. Upon insertion it is found that the un-charged pair of plates have now in fact become charged also. It here can be seen that a cube of dielectric induction can be carried through space, from one set of plates to another set of plates.

This induction is contained by the boundaries of the 10-C oil. Well golly-gee Mr. Wizard, what happened to all those electrons, isn't oil an insulator?

Here given has been various examples of dimensional relations involving space. First order space has been the long distance power line, second order space has been the package of spaghetti, third order space has been the cube of 10-C transformer oil, and, over the incessant barking of the poodle, fourth order space has been invoked as a product of conjugate flux densities.

With the understanding hereby developed, it is now possible to enter development of the concept of inductance and of capacitance, along with their use in the application of the metrical dimension of space to the substantial dimensions of electric induction. From this can be derived a substantive concept of the aether.

Transmission 9

Inductance and Capacitance

In its most general form, the basic concept of an electrical configuration in electrical engineering terms is,

1. A metallic-dielectric geometric structure,

2. A bound electric field of induction, this representing STORED ENERGY within the containing geometric structure, and

3. An exchange of electrical and mechanical forces between the electric field and the material geometric structure.

It is in statement 3 that the concepts of INDUCTANCE and of CAPACITANCE enter the electric dimensional relations. It is through the dimensional relations of inductance and capacitance that the electric field engages in the interaction with the geometry in which it is bound. It is also here that we find the most significant dimensional misrepresentations which occlude the understanding of the phenomenon of electricity.

A COMMON LANGUAGE FOR ELECTRICAL ENGINEERING

The existence of the dielectric field of induction, Psi, in Coulombs, gives rise to an electro-static potential, *e*, in Volts. Conversely an electro-static potential, *e*, in Volts, gives rise to the dielectric field, Psi, in Coulomb. It is a "chicken or egg", a matter of versor position along a cycle. Here we have a pair of dimensional relations, Psi, and, *e*, that exist in proportion to each other. It hereby follows that a proportionality factor must exist expressing the ratio of the pair of dimensional relations, Psi, in Coulomb, and *e*, in Volt. Considering the dielectric induction as a primary dimension, not a dimensional relation, then the variation of the primary dimension is with respect to the secondary dimension. Primary per secondary, Psi per *e*. An example is a package of spaghetti, spaghetti is a primary dimension, package, per square inch a secondary dimension. Hence the dimensional relation of the proportion of dielectric, Psi, in Coulombs, to the electro-static potential, *e*, in Volts.

This is then given as,

Volts equals Coulombs per Farad,

$$e = \frac{\Psi}{C} \quad \text{Volts}$$

The ratio, Psi over *e*, establishes a new dimensional relation. This relation, a factor of proportion is called the CAPACITANCE, C, in FARAD.

That is,

Farads equals Coulombs per Volt,

$$C = \frac{\Psi}{e} \quad \text{Farad}$$

If then it takes a very small magnitude of electro-static potential, *e*, to engender a very large quantity of dielectric induction, Psi, then the geometry supporting this induction is said to have a high capacitance, C. It is then called a CAPACITOR. One electro-static unit of capacitance is close to one Pico-farad, the one over *c* squared renders this 10 percent off.

Thus we can state a "Law of Dielectric Proportion", C, in Farads, is the proportion of the QUANTITY of dielectric induction, Psi, in Coulombs, to the MAGNITUDE of electro-static potential, *e*, in Volts. The Coulomb per Volt, or Farad of electro-static capacity.

It hereby follows that for a given "package", or quantity, of dielectric induction, a variation of the capacitance must give rise to a proportional variation of the electro-static potential, that is, a decrease in capacitance must give rise to an increase in electro-static potential. This is the Law of Dielectric Proportion.

The same line of reasoning follows for the magnetic field of induction. The existence of the magnetic field, Phi, in Webers, gives rise to a magnetomotive force, or M.M.F., i, in Amperes. Again, it is a versor, chicken or egg.

Here again is a pair of dimensional relations that exist in proportion to each other, Psi and i.

Thus the ratio, or factor of proportion, is given as,

Henry, or Weber per Ampere,

$$\frac{\Phi}{i} = L$$

The ratio of Psi to i results in a new dimensional relation. This factor of proportion is the dimensional relation called INDUCTANCE, L, in Henry. L equals Phi over i. L in Henry is the proportionality factor between the quantity of magnetic induction to the magnitude of the M.M.F. The Weber per Ampere, or Henry of magnetic inductance.

It then follows, for a given "package", or quantity of magnetic induction, that a variation of the inductance must give rise to a variation of the M.M.F. This is to say, a decrease in inductance must give rise to a proportional increase in current, or M.M.F. This is the Law of Magnetic Proportion.

Heretofore established is the pair of dimensional relations,

(1) The Law of Dielectric Proportion

Coulomb per Volt, or Farad, C

$$C = \frac{\Psi}{e} \quad \text{Farad}$$

(2) The Law of Magnetic Proportion

Weber per Ampere, or Henry, L

$$L = \frac{\Phi}{i} \quad \text{Henry}$$

Reduction to Primary Dimensions

In the expressions for the law of dielectric proportion, and the law of magnetic proportion, that is, the capacitance and inductance, the relations are not given entirely in primary dimensions. Both e, in Volts, and i, in Amperes, are not primary dimensions, they are secondary dimensional relations. These relations must be expanded in order to express capacitance and inductance in terms of primary dimensions only.

By the Law of Magnetic Induction,

(1) Volt, or Weber per Second

$$E = \frac{\Phi}{t} \quad \text{Volts}$$

And the Law of Dielectric Induction,

(2) Ampere, or Coulomb per Second

$$I = \frac{\Psi}{t} \quad \text{Amperes}$$

Combining terms, for the dielectric capacitance,

(3) Farad, or Coulomb per Volt

$$C = \frac{\Psi}{e} \quad \text{Farad}$$

Or by substitution,

Farad equals Coulomb-Second per Weber,

$$C = \frac{\Psi t}{\Phi} \quad \text{Farad}$$

This is the primary dimensional relation expressing capacitance, C, in Farad. Now the primary dimension of Time has re-emerged into what has been a space relation. More on this later.

It was established early on that the ratio of the total dielectric induction, Psi, to the total magnetic induction, Phi, gives rise to the dimensional relation, Y, the admittance in Siemens. By substituting the dimensions of Siemens for the ratio Coulomb per

Weber, into the expression for Farad, this gives rise to,

(4) Farad, or Coulomb-Second Per Weber

$$C = \frac{\Psi t}{\Phi} \qquad \text{Farad}$$

And gives,

(5) Farad, or Siemens-Second

$$C = Yt \qquad \text{Farad}$$

Rearrangement of terms in (5) results in an important dimensional relation,

(6) Farad per Second, or Siemens

$$\frac{C}{t} = Y \qquad \text{Siemens}$$

That is, C over T gives the dimensional relation of Siemens. This new relation is the SUSCEPTANCE, B, in Siemens.

It is hereby established that the dimensional relation of Siemens can now be expressed in two distinct forms,

(7) Admittance, Y, in Coulomb per Weber

$$Y = \frac{\Psi}{\Phi} \qquad \text{Siemens}$$

(8) Susceptance, B, in Farad per Second

$$B = \frac{C}{t} \qquad \text{Siemens}$$

More on this later.

The same considerations apply to the magnetic field of induction, and its Law of Magnetic Proportion, the inductance, L, in Henry,

(9) Weber per Ampere, or Henry

$$L = \frac{\Phi}{i} \quad \text{Henry}$$

Substituting gives,

(10) Weber-Second per Coulomb, or Henry

$$L = \frac{\Phi t}{\Psi} \quad \text{Henry}$$

And by the relation,

(11) Weber per Coulomb, or Ohm

$$\frac{\Phi}{\Psi} = Z \quad \text{Ohm}$$

It is then given,

(12) Ohm-Second, or Henry

$$L = Zt \quad \text{Henry}$$

And thus,

(13) Henry per Second, or Ohm

$$\frac{L}{t} = X \qquad \text{Ohm}$$

This hereby derived dimensional relation for Ohm, or Henry per second, is called the REACTANCE, X, in Ohm. Again, as with the Siemens, a dual dimensional relation exist with regard to the Ohm, the impedance, Z, and the reactance, X.

We here have established a new pair of dimensional relations. These relations involve a time rate of variation, this analogous to the time rate relations, the Faraday and Maxwell Laws of Induction, given again.

That is,

(A) Farad per Second or Siemens, B

$$\frac{C}{t} = B \qquad \text{Siemens}$$

(B) Henry per Second or Ohm, X

$$\frac{L}{t} = X \qquad \text{Ohm}$$

Two alternate views present themselves as to the time rate of variation. One is the condition that the capacitance and inductance in themselves are constants, time invariants, it is that the forces, electro-static potential, and magneto-motive force, are time variant. The e, and i, are in variation with respect to time. This is the condition for the relations of susceptance, B, in Siemens, and of reactance, X, in Ohms.

For example, take a one Henry inductance coil. The given line voltage is 120 volts A.C. in variation at a rate of 377 radians per second, or 60 cycles per second. Hereby the reactance of the one (1) Henry inductor is thus the product of 1 and 377 or 377 Ohms, or Henry per Second.

The application of 120 volts A.C. to this inductor hereby gives rise to a current of,

$$\frac{120}{377} \qquad \text{Ampere, or Volt per Ohm}$$

For the sake of simplicity, let us say this is about a quarter ampere, one fourth of an amp.

The product of 120 volts and one fourth amp gives the electrical activity as,

$$\frac{120}{4} \text{ Volt} - \text{Ampere or 30 Volt} - \text{Ampere reactive}$$

This is the electrical activity of the one Henry coil across 120 volt A.C. at 60 cycles.

Carrying the Law of Magnetic Proportion one step further, this one Henry inductance coil, in its windings, has 1000 passes, or turns around its core. This hereby gives rise to a M.M.F. of 1000 times one fourth ampere, or a total of 250 ampere- turns. This magnetomotive force, or compound current is developed in a one, 1, Henry coil. Hereby, by the Law of Magnetic Proportion, for a current of one quarter ampere through 1000 turns gives rise to the quantity of magnetic induction, 250 Webers.

The Variation of Inductance and Capacitance With Respect to Time

We have heretofore established a new pair of dimensional relationships. These are the magnetic inductance, L, in Henry, and the electro-static capacity, C, in Farad. Derived from these dimensional relations is a pair of electrical laws.

That is,

(I) The Law of Dielectric Proportion

The ratio of the quantity of dielectric induction, Psi in Coulomb to the magnitude of the electro-static potential, e, in Volt,

(1) Coulomb per Volt, or Farad

$$\frac{\Psi}{e} = C \qquad \text{Farad}$$

(II) The Law of Magnetic Proportion

The ratio of the quantity of magnetic induction, Phi, in Weber, to the magnitude of the M.M.F., i, in Ampere,

(2) Weber per Ampere, or Henry

$$\frac{\Phi}{i} = L \qquad \text{Henry}$$

Through algebraic rearrangement a pair of secondary dimensional relations alternately define, in a new form, the total dielectric induction, Psi, in Coulomb, and the total magnetic induction, Phi, in Weber.

For the dielectric induction,

(3) Coulomb, or Volt-Farad

$$\Psi = Ce \qquad \text{Coulomb}$$

And for the magnetic induction,

(4) Weber, or, Ampere-Henry

$$\Phi = iL \qquad \text{Weber}$$

Hence, the total dielectric induction, Psi, in Coulomb, is the product of the potential, e, in volts, and the capacitance, C, in Farads. Likewise, the total magnetic induction, Phi, in Webers, is the product of the M.M.F., *i*, in Amperes, and the inductance, *L*, in Henrys.

That is,

$$\Psi = eC$$

$$\Phi = iL$$

(Psi equals *e* times C; Phi equals *i* times L)

In the expression of the variation of the parameters which constitute the dimensional relations involving capacitance and inductance, two distinct conditions can exist. First is the capacitance and the inductance are time invariant, and the variation with respect to time resides in the relations of potential, *e*, and of M.M.F., *i*. Here derived are the susceptance and the reactance. In the alternate form of expression, it is the potential, *e*, and the M.M.F., *i*, that are time invariant, and the variation with respect to time resides in the relations of capacitance and inductance as geometric coefficients. Geometry in time variation.

In general, time invariance of L and C, or time invariance of *e* and *i* each can be considered as a limiting case. Each can be in variation with respect to time at their own individual time rates. That is, for the dielectric both C and e can be in variation, and for the magnetic both L and i can be in variation. Consider the A.C. induction motor. Here is form of magnetic inductance in which both the inductance, L, and the M.M.F., i, are in time variation, L with the rotational geometric variation, and i with the rotational cyclic variation of M.M.F. The difference between the rotational frequency of i is called the slip frequency. The rotor continuously falls behind the rotation of the magnetic field, dragging energy out of this field and delivering it to the output shaft of the motor.

Considering the pair of primary dimensional relations, it is for the dielectric induction,

(5) Farad per second, or Siemens

$$\frac{C}{t} = Y \qquad \text{Siemens}$$

And for the magnetic induction,

(6) Henry per second, or Ohm

$$\frac{L}{t} = Z \qquad \text{Ohm}$$

It is established that a distinct pair of conditions exist with regard to the variation with respect to time. Either the capacitance or inductance is in variation, or the potential or M.M.F. is in variation, with respect to time.

For the condition of time invariant L and C it is given,

(7) Farad per second, or Siemens, The Susceptance, B

$$\frac{C}{t} = B \qquad \text{Siemens}$$

(8) Henry per second, or Ohm, The Reactance, X

$$\frac{L}{t} = X \qquad \text{Ohm}$$

In the second case, the L and C are in variation with respect to time. The forces, i and e, are held constant, or time invariant. Here the variation with respect to time exists with the metallic-dielectric geometry itself. This hereby produces a variation in the geometric coefficients of capacitance or inductance.

These relations are given as,

(9) Farads per second, or Siemens, The Conductance, G

$$\frac{C}{t} = G \quad \text{Siemens}$$

(10) Henry per second, or Ohm, The Resistance, R

$$\frac{L}{t} = R \quad \text{Ohm}$$

This CONDUCTANCE, G, and this RESISTANCE, R, represent the relations derived from the time variation of capacitance and from the time variation of inductance, respectively.

It is through this form of parameter variation that the energy stored in the electrical field bounded by the geometric structure is here given to an external form. This is to say, energy is taken out of the electric field and delivered elsewhere.

For a closed system, the energy stored within the electric field is lost, or dissipated, from this system. It is then ENERGY LEAKAGE from the closed system. Considering the condition of a time invariant, or stationary geometric structure, this structure exhibiting the dissipation of the energy stored within the electric field bound by the structure, the conductance, G, and the Resistance, R, are the representations of energy leakage from the dielectric and magnetic fields respectively.

For example, consider one span of a "J carrier" open wire transmission pair. Here the conductance, G, is the "leakage conductance" of the glass telephone insulator, the resistance, R, is the "electronic resistance" of the copper weld telephone wire. These represent the energy dissipation of one span of line.

This conductance, G, represents a "molecular loss" WITHIN the glass of the insulator. This resistance, R, represents a "molecular loss" WITHIN the metal of the wire. Hence it is the molecular losses of the metallic-dielectric geometry itself that gives rise to an energy leakage from a closed system. The molecular agitation and cyclic hysteresis exist within the molecular dimensions of the physical mass of the bounding geometric structure.

These consist of a multitude of minute variations of the capacitance and inductance of the geometric form. On a microscopic level the material substance of this form is indefinite, a kind of blur in space, due to the multitude of minute variations of positions in space. These tiny motions, hereby through parameter variation, convert the energy stored in the electric field into random patterns of radiation. By experiment it can be shown that this energy leakage exists in proportion to the temperature of the material form storing energy within its bound electric field.

In general, the electrostatic potential, e, in Volt, renders the insulators hot, the magnetomotive force, i, in Ampere, renders the wires hot. Also, it is found that this heating increases with increasing frequency of the potential, e, or the M.M.F., i. It is here where the prevailing concept of the "electron" is to be found. Hence it is the motions of the electrons that give rise to the energy loss in an electrical system.

Electrons represent energy dissipation. However, the pedant, the mystic, and the dis-informer all tell us that the electron is what conveys energy, the complete opposite!

Parameter Variation Continued

In the last section the dimensional relations of conductance and resistance were developed for the condition of a static, or stationary, metallic-dielectric geometry. The conductance represents the leakage of energy from the dielectric field, and the resistance represents the leakage of energy from the magnetic field. Energy loss is internal to the physical mass which constitutes the metallic-dielectric geometry. This loss is of molecular form.

Instead of the parameter variation resulting from internal motions, there exists the parameter variation which results from external motion. This parameter variation with respect to time is the result of the contiguous parts of the geometric form being in relative motion with respect to each other. Again the A.C. induction motor serves as an example of such a geometric structure. Here is a metallic-dielectric geometric structure with relative motion between its physical parts. A motor or a generator operate through parameter variation via rotational motion. An example is the common "Electro-static" generator, such as the "Wimshurst Machine", a rotating variable electro-static condenser.

In general the metallic-dielectric geometry delivers mechanical force as an electric motor, or is driven by mechanical force as an electric generator.

Mechanical/Electrical parameter changes, these as, Farad per second and Henry per second, give rise to the metallic-dielectric geometry becoming and electric motor, taking energy from the field, or becoming an electric generator, giving energy to the field.

In the case which the geometry is taking energy as an electric motor, it is for a dielectric machine, a parametric Conductance, G, results,

(1) Farad per second equals Conductance, G, in Siemens

$$\frac{C}{t} = G \qquad \text{Siemens}$$

And for a magnetic machine, a parametric Resistance, R, results,

(2) Henry per second equals Resistance, R, in Ohms

$$\frac{L}{t} = R \qquad \text{Ohm}$$

R and G here represent the removal of energy from the electric field, just as with the condition of molecular losses.

For the condition of a mechanically driven metallic-dielectric geometry giving energy as an electric generator, an alternate form of dimensional expression is desired. These expressions serve to distinguish that part of the relations which represent the loss of energy as distinct from that part of the relations which represent the gain of energy. The square root of positive one is the "operator" which distinguishes the gain part from loss part. It is supply, or demand.

These alternate dimensional relations are, for the dielectric field,

(3) Farad per second equals the Acceptance, S, in Siemens

$$\frac{C}{t} = S \qquad \text{Siemens}$$

And for the magnetic field,

(4) Henry per second equals the Receptance, H, in Ohms

$$\frac{L}{t} = H \quad \text{Ohm}$$

Hence for the dielectric machine an ACCEPTANCE, *S*, in Siemens, and for a magnetic machine a RECEPTANCE, *H*, in Ohm. Where *R* and *G* represent energy consumption coefficients, and *S* and *H* represent energy production co-efficients.

A few observations are in order here. First, existing technology produces machines which are strictly magnetic, such as the A.C. induction motor, or strictly dielectric, such as the Wimshurst Machine. No machine is produced where the magnetic and the dielectric fields work together in an electric field. What relationship of the forces, potential, *e*, and M.M.F., *i*, gives rise to equal and opposite mechanical force, this now applied to a rotating geometry?

Second, not all parameter changes are the result of mechanical forces, nor random molecular motions. The magnetic amplifier is one such case, here a parametric inductance controlled by an auxiliary M.M.F. On the molecular level, certain plasma discharge tubes, such as the common fluorescent lighting tube, give rise to an assortment of parameter variations which can produce as well as consume energy from the electric field. Here is a vast realm for theory and experiment.

Transmission 10

The Telegraph Equation, Part One

From the previous sections, a set of dimensional relations has been established as derived from four basic electrical laws,

(1) The Law of Dielectric Proportion

(a) Farad, or Coulomb per Volt

$$C = \frac{\Psi}{e} \qquad \text{Farad}$$

(b) Coulomb, or Volt-Farad

$$\Psi = eC \qquad \text{Coulomb}$$

(2) The Law of Magnetic Proportion

(a) Henry, or Weber per Ampere

$$L = \frac{\Phi}{i} \qquad \text{Henry}$$

(b) Weber, or Ampere-Henry

$$\Phi = iL \qquad \text{Weber}$$

(3) The Law of Dielectric Induction

(a) Ampere, or Coulomb per Second

$$I = \frac{\Psi}{t} \qquad \text{Ampere}$$

(b) Coulomb, or Ampere-Second

$$\Psi = It \qquad \text{Coulomb}$$

(4) The Law of Magnetic Induction

(a) Volt, or Weber per Second

$$E = \frac{\Phi}{t} \quad \text{Volt}$$

(b) Weber, or Volt-Second

$$\Phi = Et \quad \text{Weber}$$

Recombination of these dimensional relations, or electrical laws, then expresses a pair of ratios, of primary dimensions in variation with respect to time.

Hence, for the dielectric field,

(5) Farad per Second, or Siemens

$$\frac{C}{t} = Y \quad \text{Siemens}$$

And for the magnetic field,

(6) Henry per Second, Or Ohm

$$\frac{L}{t} = Z \qquad \text{Ohm}$$

From this pair of dimensional relations are derived a series of energy transfer and storage coefficients.

Grouping these into a pair of categories, these are given as,

(I) Energy Storage and Dissipation

(a) Dielectric Energy Storage, Susceptance, B in Farad per Second or Siemens

$$B = \frac{C}{t} \qquad \text{Siemens}$$

(b) Magnetic Energy Storage, Reactance, X in Henry per Second or Ohms

$$X = \frac{L}{t} \qquad \text{Ohm}$$

And

(c) Dielectric Energy Dissipation, Conductance, G in Farads per Second or Siemens

$$G = \frac{C}{t} \quad \text{Siemens}$$

(d) Magnetic Energy Dissipation, Resistance, R in Henry per Second or Ohms

$$R = \frac{L}{t} \quad \text{Ohm}$$

(II) The Energy Consumption or Production

(a) Dielectric Energy Consumption, Conductance, G in Farad per Second or Siemens

$$G = \frac{C}{t} \quad \text{Siemens}$$

(b) Magnetic Energy Consumption, Resistance, R in Henry per Second or Ohms

$$R = \frac{L}{t} \quad \text{Ohm}$$

And

(c) Dielectric Energy Production, Acceptance, S in Farad per Second or Siemens

$$S = \frac{C}{t} \quad \text{Siemens}$$

(d) Magnetic Energy Production, Receptance, H in Henry per Second or Ohms

$$H = \frac{L}{t} \quad \text{Ohm}$$

SPECIAL NOTE: These various groupings of coefficients exist in distinct, independent, time frames. The dissipation coefficients are the result of random molecular variations, that is, noise. The consumption coefficients are harmonic in nature, relating to the operating frequencies, likewise for the production coefficients. The random and the harmonic time functions are NOT ADDITIVE. In general, the combinations of these coefficients appear as versor sums. More on this later.

Since the total electric induction is the product of the total dielectric induction and the total magnetic induction, there exists the products of the coefficients of dielectric induction and the coefficients of magnetic induction. These products give rise to a set of electrical factors. These factors, the product of the dielectric part, in Siemens, and of the magnetic part, in Ohm, gives rise to the dimensional relation.

(7) Ohm-Siemens, or Numeric

$$ZY \qquad \text{Numeric}$$

Hence, this derived dimensional relation, or FACTOR, is a numeric, that is, dimensionless. Since both the Ohm and the Siemens are versor quantities, it follows that this numeric is also a versor, a dimensionless versor magnitude. It is not a scalar, it is a versor with a position in time.

These factors are hereby established to be dimensionless versor magnitudes.

Combining the dielectric and magnetic coefficients gives the following factors,

(a) The Energy Storage Factor, XB, in Ohm-Siemens or Henry-Farad per Second Squared

$$XB = \frac{LC}{t^2} \qquad \text{Energy Storage Factor}$$

(b) The Energy Loss Factor, RG, in Ohm-Siemens or Henry-Farad per Second Squared

$$RG = \frac{LC}{t^2} \qquad \text{Energy Loss Factor}$$

And finally,

(c) The Energy Gain Factor, HS, in Ohm-Siemens or Henry-Farad per Second Squared

$$HS = \frac{LC}{t^2} \qquad \text{Energy Gain Factor}$$

Hereby it is, *HS* supplies the energy, *XB* holds the energy, *RG* removes the energy. These three factors define the movement of electricity thru the dimension of time, this for a generalized electrical configuration.

It is usually that the electrical configuration, the metallic-dielectric geometry, exhibits only energy losses, no component of energy gain exists. An example is one span of a J-Carrier open wire transmission line. This line holds energy in its bound electric field of induction, but a portion of this energy is lost through molecular action within the glass insulators and within the copper weld wires. There exists no component of energy gain in this span of open wire line. Here it is the parametric terms vanish. No factor *HS* exists and *RG* is pure dissipation. These simplifications allow for the algebraic expression in an archetypical form of the generalized electrical configuration.

Given the basic dimensional relations,

X, the Reactance, in Henry per Second or Ohms

$$X = \frac{L}{t} \quad \text{Ohm}$$

And

B, the Susceptance, in Farad per Second or Siemens

$$B = \frac{C}{t} \quad \text{Siemens}$$

These relations representing energy exchange between the dielectric field, and the magnetic field, of inductions. This energy exchange is in an alternating form.

It is also,

R, the Resistance, in Ohms

And

G, the Conductance, in Siemens

These relations representing energy removal from the magnetic field, and the dielectric field, of inductions. This energy loss is in a continuous form. Hereby XB is the alternating "current" factor, and RG is the direct, or continuous, "current" factor.

Obviously, in the situation of an electric generator, *HS* could replace *RG* in such a configuration. Here energy is produced in a manner of negligible losses, and thus *RG* drops out of the equation. It is however, a system or configuration exhibiting both loss and gain and requires a more complex algebraic expression. This is developed in the final section of "Symbolic Representation of the Generalized Electric Wave" by E. P. Dollard.

Combining terms with like dimensional relations, that is, Ohm and Henry per Second, or Siemens and Farad per Second, gives rise to a total impedance, or a total admittance of the electrical configuration.

Hence it is,

(I) The Total Admittance, *Y*, in Siemens

$$(8) \quad Y = G - jB \qquad \text{Siemens}$$

The versor sum of the Conductance, *G*, in Siemens, and the Susceptance, *B*, in Farad per Second.

(II) The Total Impedance, *Z*, in Ohms

(9) $Z = R + jX$ **Ohm**

The versor sum of the Resistance, R, in Ohms, and the Reactance, X, in Henry per Second. Here Y represents the dielectric field, and Z represents the magnetic field. The electric field is the product of the dielectric field, and the magnetic field. Q is Psi times Phi. Taking then the product of the total dielectric Admittance, Y, in Siemens, and the total magnetic Impedance, Z, in Ohm, gives the dimensional relation.

(10) Siemens-Ohm or Numeric

ZY Numeric

Hence ZY is a dimensionless magnitude, it having a versor position in time, since both Z and Y have a versor position in time. The product of the two versors is also a versor. ZY is not scalar, it is a dimensionless versor magnitude. It represents a wave propagation in the dimension of time, a TIME WAVE.

The "Telegraph Equation", Finale'

It has been given that the product ZY, in Siemens-Ohm, is a dimensionless magnitude, having a versor position in time.

It is the product of a pair of versor sums,

$$Y = G - jB \quad \text{Siemens}$$

$$Z = R + jX \quad \text{Ohm}$$

However, the product of versor sums are also versor sums. Taking this product,

$$ZY = (R + jX)(G - jB) \quad \text{Numeric}$$

And factoring like terms, gives the following factors,

(I) The POWER FACTOR

$$a = (XB + RG)$$

Where *a* is the power factor, the percent of energy lost from the total movement of electrical energy in an electrical configuration.

(II) The INDUCTION FACTOR

$$b = (XG + RB)$$

Where *b* is the induction factor, the percent of energy stored by the total movement of energy in an electrical configuration.

Here, the resulting sub-factors are defined,

For the power factor,

- *XB*, the factor representing the cyclic exchange of energy between dielectric and magnetic forms.
- *RG*, the factor representing the acyclic dissipation of energy from both dielectric and magnetic forms.

And, for the induction factor,

- *XG*, the factor representing the transfer of energy out of magnetic form and into dielectric form.

- **RB**, the factor representing the transfer of energy out of dielectric form and into magnetic form.

The propagation constant, ZY, in Ohm-Siemens can hereby be expressed as the versor sum of the power factor, *a*, in percent, and the induction factor, *b*, in percent.

This results in,

$$ZY = (ha + jb)$$

Ohm-Siemens, or Total Percent

ZY must always equal to 100 percent, but it has a variable position in time, this expressed as a resultant of *ha* and *jb*.

The versor operators are defined as,

h, the roots of the square root of positive one

j, the roots of the square root of negative one

Expanding the expression for the propagation constant, ZY, as a versor sum of the expressions for a and for b, gives,

$$ZY = h(XB + RG) + j(XG - RB)$$

Hereby established is the most important algebraic expression of dimensional relations, this defining the movement in time of electrical energy in any electrical configuration.

This algebraic expression is called the **Heaviside Telegraph Equation**. It is in this expression the electrical energy is expressed directly in its four pole archetype. Note that this four polar archetype underlies all Native American art forms. Is this related to America as the birthplace of electrical technology through Tesla, Edison, and Steinmetz. Europe was too consumed in self-edification mathematics, except for the work of GÖTHE. This algebraic expression gives a pair of waves in motion through the dimension of time, one moving forward in time, the other backward in time.

[Note: See "Theory and Calculation of Transient Electric Phenomena", C. P. Steinmetz, the chapter, "Resistance, Inductance, and Capacity".] Here **(R + S)** is forward in time and **(R - S)** is backwards in time.

Geometrically, this expression represents a pair of counter propagating logarithmic spirals. This spiral form is demonstrated in Ernst Guillemin, "Communication Networks" Volume One, and in "Theory and Calculation of A.C. Phenomena", Appendix – "Oscillating Currents", by C. P. Steinmetz. It is important to remember that these "motions in time" are of a versor form, finding no equivalence in spatial representation except by analogy. There is no such thing as a surface of time, no 2D time. These are versor, not vector expressions.

Expressing the four distinct sub-factors, the versor combination of which gives the propagation constant, ZY, in Ohm-Siemens, it is,

(I) The Power Factor Pair

- XB, the "axial" product, the longitudinal component of energy motion in time, forward and backward in an alternating manner.

- RG, the "dot" product, the scalar component of energy dissipation, this independent of time.

(II) The Induction Factor Pair

A COMMON LANGUAGE FOR ELECTRICAL ENGINEERING

- XG, the "cross" product, the transverse component energy transfer thru time from
- magnetic to dielectric. This is a clockwise versor around axis XB.
- RB, the "cross" product, the transverse component of energy transfer thru time from dielectric to magnetic. This is a counter-clockwise versor around axis XB.

In the pedantic, mystic, and disinformation world, there are two products, the dot and the cross. Here exists four products, axial, dot, and a conjugate pair of cross products. Here is why misunderstanding exists, the basis for the "longitudinal scalar" idiots.

In its versor form, the Telegraph Equation is expressed symbolically as,

$$k(ZY) = ha + jb$$

Where the magnitude, ZY, represents the electricity, and the operator, k, represents its versor position in time. This is given for a 360 degree scale on a power factor meter, an analog computer for the expression of $k(ZY)$. (ZY is the pointer, k is the scale, that simple.)

Expressing a versor relation as,

$$k = jh$$

That is, negative one to the one half power times positive one to the one half power, gives negative one to the one fourth power,

$$-1^{\frac{1}{2}} \cdot 1^{\frac{1}{2}} = -1^{\frac{1}{4}}$$

Since this "fourth root" of negative one suggests a conjugate of the fourth root of positive one, it is then,

k, the roots of the EIGHTH root of positive one, $\sqrt[8]{+1}$

This as the most general versor operator for the Telegraph Equation.

Conclusion

Here we are beyond the scope of this elementary series of discussions on the rudiments of electrical theory. This series of transmissions concludes here. But it was fun, don't you think?

For a more in-depth study of this subject, see the following:

"Theory and Calculation of A.C. Phenomena", C. P. Steinmetz, the chapters, "Power and Double Frequency Quantities" and the appendix "Roots of the Unit".

"Symbolic Representation" Papers by E. P. Dollard, and all references given in these papers.

"Electro-Magnetic Theory", O. Heaviside, in particular the development of his "Telegraph Equation".

"Physics and Mathematics in Electrical Communication", James Owen Perrine.

Finally, for an excellent musical portrayal of the ZY relationship listen to G. F. Handel, "Alexander's Feast, or the Power of Music", the final choral movement. It is a good ending to this series of writings.

A COMMON LANGUAGE FOR ELECTRICAL ENGINEERING

List of Symbols

Category		Symbol	Name	Description
Primary Dimensions	Substantial	Q	Planck	Unit of Electric Induction
	Metrical	l	Centimeter	Unit of Space
	Metrical	t	Second	Unit of Time
Derived Primary Dimensions	Substantial	Φ	Weber	Line of Magnetic Force
	Substantial	Ψ	Coulomb	Line of Dielectric Force
Secondary Relations	Substantial & Metrical	W	Joule	Unit of Energy
	Substantial & Metrical	P	Watt	Unit of Power
	Substantial & Metrical	E	Volt	Electromotive Force
	Substantial & Metrical	I	Ampere	Displacement Current
	Substantial & Metrical	e	Volt	Electrostatic Potential
	Substantial & Metrical	i	Ampere	Magnetomotive Force
	Substantial & Metrical	Y	Siemens	Admittance
	Substantial & Metrical	Z	Ohm	Impedance
	Substantial & Metrical	G	Siemens	Conductance
	Substantial & Metrical	R	Ohm	Resistance
	Substantial & Metrical	B	Siemens	Susceptance
	Substantial & Metrical	X	Ohm	Reactance
	Substantial & Metrical	S	Siemens	Acceptance
	Substantial & Metrical	H	Ohm	Receptance
	Substantial & Metrical	d	N/A	Electrostatic Gradient
	Substantial & Metrical	m	N/A	Magnetomotive Force Gradient
	Substantial & Metrical	d'	N/A	Electromotive Force Gradient
	Substantial & Metrical	m'	N/A	Displacement Current Gradient
	Metrical	C	Farad	Capacitance
	Metrical	L	Henry	Inductance
Versor Symbols	N/A	h	Versor Operator	Roots of SQRT (+1)
	N/A	j	Versor Operator	Roots of SQRT (-1)
	N/A	k	Versor Operator	Roots of Product of jk
Complex Frequency	Metrical	v	Neper-Radians per Second	Decibel Cycles per Second
Power & Induction Factor	N/A	a	Percent	Power Factor
	N/A	b	Percent	Induction Factor

Recommended Reading

During the course of reading this book, you will find the author making various recommendations for further study.

These include, but are not limited to:

Previous works by the author, Eric P. Dollard

As well as works by:

Charles P. Steinmetz
Oliver Heaviside
Nikola Tesla

And the following mathematicians:

Grassman
Hamilton
Tait
MacFarlane

As you read, make note of these recommendations and follow-up on them with later study.

Glossary

All good text books have a glossary at the end. What follows is a list of all the terms you would want in a glossary of this book, along with a space for YOU to write in their definitions. Consider this exercise your final exam in the author's course on Electricity, 101, Definitions of Terms and Conditions.

Admittance:

Aether:

Ampere:

Capacitance:

Conduction current:

Conductor:

Coordinates:

Coulomb:

Counter-space:

Dielectricity:

Dimensions:

Displacement current:

Electricity:

Electro-Motive Force:

Energy:

Farad:

Frequency:

Hysteresis:

Henry:

Impedance:

Inductance:

Induction Factor:

Insulator:

Joule:

Magnetism:

Magneto-Motive Force:

Neper-Radians per Second:

Ohm:

PHI:

Planck:

Power:

Power Factor:

PSI:

Reactance:

Siemens:

Space:

Susceptance:

Time:

Versor Operator:

Volt:

Weber:

There is a little extra room left here in case you find a few more terms you would like to define for yourself.

When you fully understand the definition of these terms, and their spatial and temporal relation to each other, you will have learned the material and graduated to the level of being able to speak intelligently and accurately about electrical engineering subjects. Congratulations.

A COMMON LANGUAGE FOR ELECTRICAL ENGINEERING

LEARN MORE ABOUT THE LEGENDARY WORK OF ERIC P. DOLLARD

Please Support Eric Dollard & EPD Laboratories, Inc., a 501(c)3 non-profit corporation by making a tax deductible donation at:
http://ericpdollard.com

ERIC DOLLARD'S OFFICIAL FORUM

http://www.energeticforum.com/eric-dollard-official-forum

A COMMON LANGUAGE FOR ELECTRICAL ENGINEERING

ENERGY SCIENCE & TECHNOLOGY CONFERENCE

Meet Eric P. Dollard in Person & Other Pioneers of the Modern-Day Tesla Movement

An Exclusive Annual Event in Hayden, Idaho, United States Only 50 Minutes from Spokane

SEATING IS LIMITED TO 150 REGISTER EARLY TO RESERVE YOUR SEAT!

http://energyscienceconference.com

A COMMON LANGUAGE FOR ELECTRICAL ENGINEERING

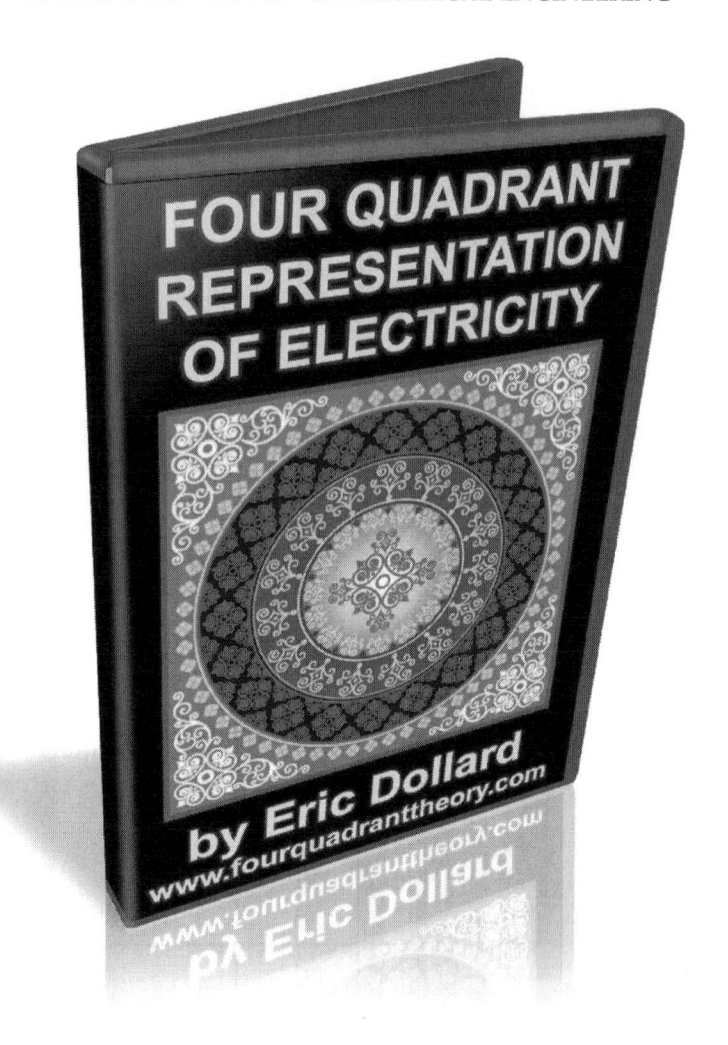

FOUR QUADRANT REPRESENTATION OF ELECTRICITY

http://fourquadranttheory.com

A COMMON LANGUAGE FOR ELECTRICAL ENGINEERING

VERSOR ALGEBRA AS APPLIED TO POLYPHASE POWER SYSTEMS

http://versoralgebra.com

A COMMON LANGUAGE FOR ELECTRICAL ENGINEERING

THE EXTRALUMINAL TRANSMISSION SYSTEMS OF TESLA AND ALEXANDERSON

http://extraluminaltransmission.com

A COMMON LANGUAGE FOR ELECTRICAL ENGINEERING

WIRELESS GIANT OF THE PACIFIC

http://wirelessgiantofthepacific.com

A COMMON LANGUAGE FOR ELECTRICAL ENGINEERING

CRYSTAL RADIO INITIATIVE

http://crystalradioinitiative.com

A COMMON LANGUAGE FOR ELECTRICAL ENGINEERING

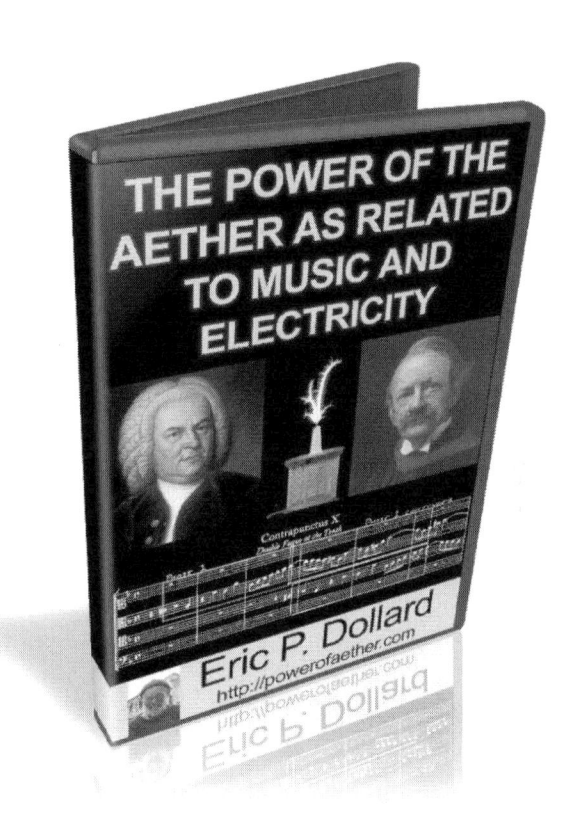

THE POWER OF THE AETHER AS RELATED TO MUSIC AND ELECTRICITY

http://powerofaether.com

A COMMON LANGUAGE FOR ELECTRICAL ENGINEERING

A & P Electronic Media
Leading Digital Publishers of Advanced Energy Science
http://emediapress.com

A portion of the proceeds from sales of Eric Dollard's books & videos go to EPD Laboratories, Inc., a 501(c)3 non-profit corporation to support his work in advancing the electrical sciences.

For More Books & Videos from the Leading Publishing Authority on the Tesla Sciences, Visit http://emediapress.com

A COMMON LANGUAGE FOR ELECTRICAL ENGINEERING

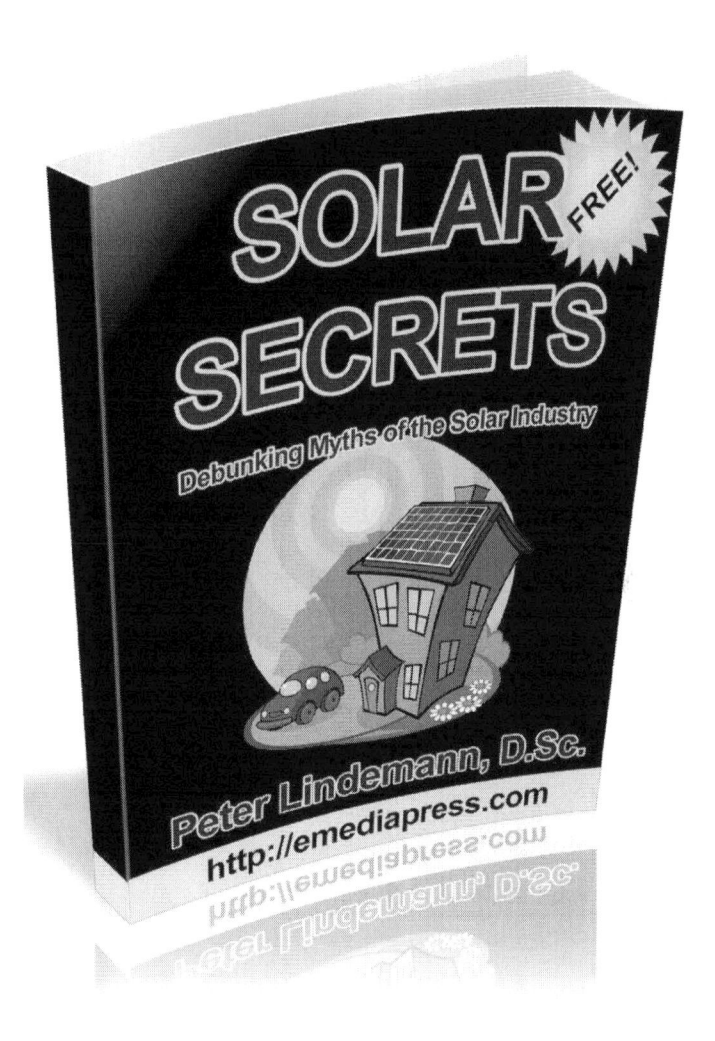

SOLAR SECRETS

FREE DOWNLOAD RIGHT NOW!

http://freesolarsecrets.com

A COMMON LANGUAGE FOR ELECTRICAL ENGINEERING

WATCH FREE VIDEOS
PRESENTED BY
TESLA MEDIA NETWORK
ON CONNECTED TELEVISION

MANY HOURS OF
FOUNDATIONAL
PRESENTATIONS BY THE
TELSA MASTERS

http://teslamedianetwork.com/

A COMMON LANGUAGE FOR ELECTRICAL ENGINEERING

Printed in Great Britain
by Amazon.co.uk, Ltd.,
Marston Gate.